Corrosion
and Related Aspects of Materials for
Potable Water Supplies

Edited by
P. McIntyre and A. D. Mercer

Proceedings of a Conference held at
The Society of Chemical Industry, London, UK, 8–9 December, 1992

The Institute of Materials

1993

Book Number 561
Published in 1993 by The Institute of Materials
1 Carlton House Terrace, London SW1Y 5DB

© 1993 The Institute of Materials

All rights reserved

British Library Cataloguing in Publication Data
Available on request

Library of Congress Cataloging in Publication Data
Available on application

ISBN 0-901716-47-2

Colour origination by Colorhouse, Sherborne, Dorset

Cover graphics and production by
PicA Publishing Services, Drayton, Nr Abingdon, Oxon

Made and printed in Great Britain

Contents

Introduction   v
W L Mercer

Drinking Water Standards and Quality   1

1 Water Quality Standards   2
A H Goodman

2 Developments in CEN in Relation to Effects of Materials on Drinking Water Quality   11
M Fielding

Materials   17

3 Corrosion of Cast Iron in Potable Water Service   18
G Gedge

4 Plastics   29
J M Marshall

5 Concrete: Its Uses in Potable Water Supplies and Interactions with Aqueous Environments   45
J Figg

6 Corrosion of Galvanised Steel in Potable Water Supplies   54
C-L Kruse

7 Copper   65
J L Nuttall

8 Lead: A Source of Contamination of Tap Water   84
R Gregory

9 Stainless Steels for Potable Water Systems   94
D Dulieu, B V Lee, M O Lewus and K W Tupholme

Users Experience   113

10 Avoidance of Corrosion Problems   114
A K Tiller

11 Effects of Water Composition and Operating Conditions on the Corrosion Behaviour of Copper in Potable Water   122
H Siedlarek, D Wagner, M Kropp, B Füssinger, I Hänssel and W R Fischer

19 Observations on the Corrosion Behaviour of Copper Pipes in Glasgow Tap Water   140
T Hodgkiess and J Akhtar

13 Influence of Operating Conditions on Materials and Water Quality in Drinking Water Distribution Systems   161
I Wagner

14 Assessment of Corrosion Trends in a Potable Water Distribution System Including an Automatic Analysis Technique   172
A M Dean and T Koch

15 Influence of Network Disinfection on Corrosion of Galvanised Steel Pipes (Use of Hypochlorite, Permanganate or Hydrogen Peroxide) — Case of Paris   195
P Leroy

16 Corrosion of a Welded Steel Pipe for a Potable Water Supply   207
P L Bonora and L Ghirardelli

17 An Unusual Form of Microbially Induced Corrosion in Copper Water Pipes   222
H S Campbell, A H L Chamberlain and P J Angell

18 Corrosion in Potable Water Systems — The Situation in Finland   232
T Kaunisto

12 A User's Point of View   240
D W Lackington

20 Study of the Influence of Alkalinity and Calcium on Copper Corrosion   259
T Hedberg and E L Johansson

21 Perforations of Polypropylene Pipes in Potable Water Caused by Cracking — A Case Study   268
W R Fischer, D Wagner, H Siedlarek and J A G C Sequeira

Index   278

Introduction

W. L. MERCER

(President of the Institute of Metals, 1990–1991)

A reliable and continuous supply of wholesome water is vital to our way of life, indeed to life itself — without water life is not possible. And yet, today, we take it for granted. It flows readily from the tap, always reliable both in respect of quantity and quality.

Water is free — or so many believe. Emotions run high when the costs of collecting, storing, purifying, distributing water and disposing of the resultant effluent are imposed through regular and ever-increasing charges. The scale of the activity is immense and the remarkable fact is the low cost within which this vital and ever reliable supply is maintained.

The Romans recognised the importance of water in civilised living. Stone aqueducts, many standing to this day, remain a tribute to their foresightedness and engineering skills. They were perhaps less fortunate in not having our understanding of the effect of various metallic contaminants. Water is a powerful solvent and many common metals it comes into contact with can be harmful to the human body. Lead vessels were widely utilised by the Romans but, being relatively expensive compared with clay, were usually restricted to the nobility and more wealthy families of the time. Thus, and particularly in the case of acidic (soft) waters and wines, lead was ingested on a regular and continuous basis. It has been claimed that, amongst other effects, this severely lowered the fertility of the more gifted members of society and, as a result, contributed to the fall of the Roman Empire!

In the UK, the Victorians were the great water engineers. Their magnificent works probably played a greater role in ensuring public health and the quality of life than any other act of man throughout history. Great dams, purification beds, major pipelines and sophisticated sewage collection systems were products of this era. The work continues to this day. The demand for water is unceasing, although the recessions of recent years coupled with improved manufacturing efficiency have significantly decreased industrial demand. This is not true of domestic demand, where with the rapid growth of labour saving aides, water use per head is still rising. New methods and new materials are meeting the challenge. Metallic and cementitious distribution systems of yesteryear are giving way to plastics which offer huge advantages in terms of lower costs, freedom from corrosion and chemical and biological inertness.

This Conference brought together many of these factors. The papers presented not only examine and update the reader on the performance of various materials in contact with potable water, but they also introduce the all important question of Standards — both national and international. In this respect, it was pleasing that the Conference attracted many overseas delegates. Almost one third came from continental Europe and some travelled from the other side of the globe.

Good clean water is vital to all. The Conference played a valuable role in setting the scene. The collected papers are a valuable reference source and statement of the current position. As such they provide a basis for continuing progress and they point the way ahead to the promise of even greater achievements in the future.

Drinking Water Standards and Quality

Water Quality Standards

A. H. GOODMAN

Drinking Water Inspectorate, Department of the Environment, London, UK

1. BACKGROUND AND INTRODUCTION

Water is often described as the universal solvent, and because it has solvent properties it can give rise to problems as it is conveyed through distribution systems to houses. For many years the only requirement for public water supply quality was that it was wholesome — a useful word but rarely defined.

In most of the early public health legislation there was an obligation on the owner of a water supply to provide wholesome water. In the Public Health Acts from over a century ago, the responsibility for ensuring that the water received by consumers was wholesome, was vested on the Medical Officer of Health. The Water Act 1945 made it the responsibility of the water company or water supplier to supply wholesome water, but again there was no definition of quality which would satisfy this requirement. As long as supplies were consumed locally, the local population became acclimatised to whatever quality it was, and only gross contamination would cause ill effects in consumers. However, as populations became more mobile, any immunity gained in one area would not necessarily provide protection to health in another, and thoughts were given to producing a minimum standard of quality of water. Microbiologically, there were such standards; for "Bacteriological Standards for Water Quality", Report 71 of the Medical Reports Series produced by the Department of Health, had produced these over 60 years ago. However, it was not easy to produce chemical standards, largely because water supplies in the UK are derived from three types of sources. In England and Wales roughly one third of supplies are drawn from groundwater sources, one third from upland catchments (and stored in upland reservoirs) and one third are drawn from lowland rivers which are used also to carry away surface water and sewage effluents from urban areas. These three types of source give different qualities of supplies, with a wide range of characteristics.

It was not only in the United Kingdom that thoughts had turned to water quality criteria. In California with a big demand for water and few resources, a publication called "Water Quality Criteria", was established in the late 1950s for drinking water, water for irrigation etc.. The importance of these Californian standards is that many of the parameters and the values attached to them were adopted elsewhere, as in the Ontario Water Quality Criteria, and in the World Health Organisation guidelines laid down in Standards for Drinking Water in the late 1950s and again in the European Standards and International Standards for Drinking Water in 1970 and 1971. When the Council of Europe and the European Commission discussed standards for river water crossing international boundaries, references were made to the WHO Standards, and the EC Directive concerning the Quality of Surface Water Intended for Abstraction for Drinking Water, 75/70/EEC, drew heavily on them. The later Directive relating to the Quality of Water for Human Consumption, 80/778/EEC, was made to be consistent with the earlier Directive.

The values given in the Californian Water Quality Criteria were not scientifically based. Some of those involved in this production admitted (e.g. Prof. Hillel Shuval at an Anglo–Israeli Seminar) that the values were agreed on pragmatic grounds from experience of what obtained in practice, or what could be measured accurately at the time. The lists of parameters in the Directives were derived similarly and included some elements or tests that were used in one or other member states, but not all. Nevertheless, it was imperative that member states implemented these Directives, and while they could make some requirements more stringent, they could not be made less so.

2. THE WATER ACT OF 1989

In England and Wales, the Water Act, 1989, gave powers to the Secretary of State for the Environment and the Secretary of State for Wales to make Regulations, known as the Water Supply (Water Quality) Regulations, 1989, in which the parameters listed in Directive 80/778/EEC were adopted(see Tables A–E), together with requirements for water suppliers to undertake sampling at frequencies appropriate to the size of the supply being tested, for which tables of minimum frequencies were given, and instructions given on determination of sampling points, where samples should be obtained to allow for securing of results of analysis of those samples to be reasonably representative of the water supplied to the zone being sampled. Similar Regulations have been made in Scotland and are being produced for Northern Ireland.

Potable Water

SCHEDULE 2

Regulation 3

PRESCRIBED CONCENTRATIONS OR VALUES

TABLE A

Item	Parameters	Units of Measurement	Concentration or Value (maximum unless otherwise stated)
1.	Colour	mg/l Pt/Co scale	20
2.	Turbidity (including suspended solids)	Formazin turbidity units	4
3.	Odour (including hydrogen sulphide)	Dilution number	3 at 25°C
4.	Taste	Dilution number	3 at 25°C
5.	Temperature	°C	25
6.	Hydrogen ion	pH value	9.5 5.5 (minimum)
7.	Sulphate	mg SO_4/l	250
8.	Magnesium	mg Mg/l	50
9.	Sodium	mg Na/l	150(i)
10.	Potassium	mg K/l	12
11.	Dry residues	mg/l	1500 (after drying at 180°C)
12.	Nitrate	mg NO_3/l	50
13.	Nitrite	mg NO_2/l	0.1
14.	Ammonium (ammonia and ammonium ions)	mg NH_4/l	0.5
15.	Kjeldahl nitrogen	mg N/l	1
16.	Oxidizability (permanganate value)	mg O_2/l	5
17.	Total organic carbon	mg C/l	No significant increase over that normally observed
18.	Dissolved or emulsified hydrocarbons (after extraction with petroleum ether); mineral oils	µg/l	10
19.	Phenols	µg C_6H_5OH/l	0.5
20.	Surfactants	µg/l (as lauryl sulphate)	200
21.	Aluminium	µg Al/l	200
22.	Iron	µg Fe/l	200
23.	Manganese	µg Mn/l	50
24.	Copper	µg Cu/l	3000
25.	Zinc	µg Zn/l	5000
26.	Phosphorus	µg P/l	2200
27.	Fluoride	µg F/l	1500
28.	Silver	µg Ag/l	10(ii)

Note (i) See regulation 3(5).
    (ii) If silver is used in a water treatment process, 80 may be substituted for 10.

TABLE B

Item	Parameters	Units of Measurement	Maximum Concentration
1.	Arsenic	µg As/l	50
2.	Cadmium	µg Cd/l	5
3.	Cyanide	µg CN/l	50
4.	Chromium	µg Cr/l	50
5.	Mercury	µg Hg/l	1
6.	Nickel	µg Ni/l	50
7.	Lead	µg Pb/l	50
8.	Antimony	µg Sb/l	10
9.	Selenium	µg Se/l	10
10.	Pesticides and related products:		
	(a) individual substances	µg/l	0.1
	(b) total substances(i)	µg/l	0.5
11.	Polycyclic aromatic hydrocarbons(ii)	µg/l	0.2

Notes (i) The sum of the detected concentrations of individual substances.
(ii) The sum of the detected concentrations of fluoranthene, benzo 3,4 fluoranthene, benzo 11,12 fluoranthene, benzo 3,4 pyrene, benzo 1,12 perylene and indeno (1,2,3-cd) pyrene.

TABLE C

Item	Parameters	Units of Measurement	Maximum Concentration
1.	Total coliforms	number/100 ml	0(i)
2.	Faecal coliforms	number/100 ml	0
3.	Faecal streptococci	number/100 ml	0
4.	Sulphite-reducing clostridia	number/20 ml	$\leqslant 1$(ii)
5.	Colony counts	number/1 ml at 22°C or 37°C	No significant increase over that normally observed

Notes (i) See regulation 3(6).
(ii) Analysis by multiple tube method.

TABLE D(i)

Item	Parameters	Units of Measurement	Maximum Concentration or Value
1.	Conductivity	µS/cm	1500 at 20°C
2.	Chloride	mg Cl/l	400
3.	Calcium	mg Ca/l	250
4.	Substances extractable in chloroform	mg/l dry residue	1
5.	Boron	µg B/l	2000
6.	Barium	µg Ba/l	1000
7.	Benzo 3,4 pyrene	ng/l	10
8.	Tetrachloromethane	µg/l	3
9.	Trichloroethene	µg/l	30
10.	Tetrachloroethene	µg/l	10

Note: (i) See regulation 3(3)(d).

TABLE E

Item	Parameters	Units of Measurement	Minimum Concentration(i)
1.	Total hardness	mg Ca/l	60
2.	Alkalinity	mg HCO$_3$/l	30

Note: (i) See regulation 3(2).

For the metals copper, zinc and lead, and for at least half of the samples taken for microbiological examination, the samples shall be taken at random in the distribution system, that is from consumers premises. The prescribed values for copper and zinc allow for moderate contamination of the samples because of corrosion of those metals where they are present, but lead has a more stringent value in the Regulations than in the Directive (50 μgL^{-1} against 100 μgL^{-1}).

3. MICROPOLLUTANTS: THE WATER SUPPLY (WATER QUALITY) REGULATIONS

While the prescribed values for pH and conductivity and minerals content go some way to prevent excessive attack on materials by the water (and, by so doing, limit the effect of materials on the quality of water) the standards were not intended for that purpose. Within the ranges of the prescribed values attached to the parameters, it is possible to find compliance for all the three major types of water in England and Wales, i.e. hard groundwaters, soft peaty coloured upland waters, and eutrophic lowland river waters, and a consumer travelling through areas deriving water supplies from these differing sources, ought not to suffer ill-effects if those supplies comply with the standards in the Regulations.

These ill-effects would be acute and easily recognised. What has for some time concerned those responsible for protection of public health is the effect of micropollutants consumed over long periods, even up to a lifetime. These micropollutants are usually suspected of bringing about chronic damage to the central nervous system or of inducing malignancy. Because of the long periods of induction before any ill-effects can be seen, epidemiological studies have little relevance, and tests on animals and animal models have to be relied on.

Over the past 15 years The Department of the Environment funded research into the occurrence of micropollutants of industrial origin which could be detected in raw waters, the effects of water treatment on them, particularly the product of reaction with disinfecting agents, and by using the Ames' Test, or modifications of it, had tested many extracts of drinking water for mutagenic activity as a screening test, or surrogate test, for carcinogenicity. Some animal tests were also conducted, and an assessment of risks from treated drinking waters was made. This potential risk to health from micro pollutants derived from materials in contact with water led to another requirement, Section 25 in the Water Supply (Water Quality) Regulations (see below):

Application and introduction of substances and products
25.—(1) A water undertaker shall not, otherwise than for the purposes of testing or research, apply any substance or product to, or introduce any substance or product into, water which is to be supplied for drinking, washing and cooking unless—

(a) the Secretary of State has for the time being approved the application or introduction of that substance or product and it is applied or introduced in accordance with any conditions attaching to that approval; or

(b) the undertaker is satisfied that the substance or product either alone or in combination with any other substance or product in the water is unlikely to affect the quality of the water supplied; or

(c) the undertaker can demonstrate that the substance or product has during the period of twelve months preceding the making of these Regulations been applied or introduced (otherwise than for the purposes of testing or research) by a water authority or a statutory water company into water supplied by it for domestic purposes; or

(d) the substance or product—

(i) was at any time before the commencement of these Regulations listed in the 15th Statement of the Committee on Chemicals and Materials of Construction for Use in Public Water Supply and Swimming Pools(**a**) or in any supplement to that Statement issued before the making of these Regulations; and

(ii) is applied or introduced in accordance with any conditions referred to in that Statement or any supplement so issued or any such conditions as varied under paragraph (5) of this regulation and any conditions imposed under that paragraph.

(**a**) The 15th Statement was issued in March 1989. Copies of it and of any supplement to bring information up to date may be obtained from the Drinking Water Inspectorate of the Department of the Environment, Romney House, 43 Marsham Street, London SW1P 3PY.

Sub-paragraphs (b) to (d) have effect subject to paragraph (4) below.

(2) An application for such an approval as is mentioned in paragraph (1)(a) may be made by any person.

(3) The Secretary of State may, if he decides to issue an approval for the purpose of paragraph (1)(a), include in the approval such conditions as he considers appropriate and, subject to paragraph (6), may at any time revoke or vary any approval he has previously given.

(4) The Secretary of State may by notice given in writing to any water undertaker prohibit it for such period as is specified in the notice from applying to, or introducing into, water intended to be supplied for drinking, washing and cooking any substance or product which the undertaker would otherwise be authorised to apply or introduce by paragraph (1)(b), (c) or (d).

(5) The Secretary of State may by notice in writing to water undertakers vary any condition contained in the 15th Statement or any supplement referred to in paragraph (1)(d)(i) or impose conditions as to the application or introduction of any substance or product listed in that Statement or any supplement.

(6) The Secretary of State may —

(a) revoke by an instrument in writing any approval given by him for the purposes of paragraph (1)(a);

(b) modify any such approval by an instrument in writing by including conditions, or varying existing conditions;

(c) issue any such notice as is mentioned in paragraph (4):

but, unless he is satisfied that it is necessary to do so in the interests of public health without notice, shall not do any of those things without giving all such persons as are, in his opinion, likely to be affected by the revocation or modification of the approval or by the issue of the notice at least six months' notice in writing of his intention.

(7) Notice shall be given forthwith by the Secretary of State to all persons likely to be affected by the making of such an instrument as is mentioned in paragraph (6)(a) or (b).

(8) At least once in each year beginning with the year 1990, the Secretary of State shall issue a list of all the substances and products in relation to which —

- (a) an approval for the purposes of paragraph (1)(a) has been granted or refused;
- (b) such an approval has been revoked or modified;
- (c) a notice has been issued under paragraph (4), with particulars of the action taken.

The Secretary of State is now advised by the Committee on Chemicals and Materials for Construction for use with Public Water Supplies and Swimming Pools. This had been set up in 1967 to look originally at synthetic polyelectrolytes being introduced into water treatment, but it extended its interests into materials of construction. It allowed manufacturers or suppliers to make voluntary applications to it for approval for use with public water, and it required the applicant to provide details of the composition (in strict confidence), the results of leaching and other tests, which had to be conducted at the applicant's expense. The Approvals were published in the Committee's 15th Statement. Now the voluntary system has been superseded by the statutory obligation of Section 25. There is confusion over the role of this Committee and that of the approvals schemes run by the Water Byelaws Advisory Service (WBAS) which requires all substances and products to be mechanically tested and examined for demonstration of fitness for use, and requires results of those tests in British Standard BS 6920: Tests for Materials in Contact with Water. These cover tests of extracts to determine leaching of metals, of organic substances, taste and odour, and extent of ability to support microbial growths. The Water Byelaws Advisory Scheme is concerned with ensuring that the Water Byelaws are complied with by those products in the consumers premises, that is, within the curtilage of the property and beyond the boundary stop-tap, and it is the responsibility of the water supplier to see that the Byelaws are implemented. The Secretary of State's approval (and the former DoE approvals scheme) is concerned only with the materials of construction used by the water supplier in the treatment works and in distribution including service reservoirs and mains and valves etc. Because of the long expected life of this equipment (about 40 years or more) the Secretary of State requires to have reassurance that over a longer period than is considered by the WBAS scheme, constituents or products of chemicals degradation from materials used in the water suppliers' equipment will not pose any perceived risk to health. The Committee on Chemicals and Materials for Construction for use with Public Water Supplies and Swimming Pools relies heavily on medical advice from the Department of Health for assessment of toxicological data on composition of products and the results of leaching tests, before it can make its recommendations to the Secretary of State to approve a substance. It is not always appreciated that the minor constituents of a product may give rise to the most concern, often because they remain unchanged in polymerisation reactions or during a curing process. Thus, plasticiser, antioxidants, colours or pigments or stabilisers are important, as well as the major ingredients. Polymeric substances are not always cured fully, so one or the other monomers may remain to be leached out. This may be so particularly where *in situ* rehabilitation of pipelines is being carried out. The quality of water in contact with the material may have a bearing on the leaching, as hard well waters may quickly form a protective scale over a surface, while lowland waters may be relatively heavily dosed with chlorine that can attack the material's surface. Thus the Committee may require materials to be examined in chlorinated water. Highly coloured upland waters, unless they have been comprehensively treated to remove the peaty colour, may deposit organically rich deposits with propensity for an acid reaction. Such deposits usually contain iron and manganese. Also waters high in nutrients, organic, nitrogen-containing and available phosphate, may produce conditions which allow thick biological films to develop, under which anaerobic conditions may occur, or acidity be produced as they decay. All waters will produce a biofilm on a surface, because although a drinking water has to be free from pathogenic organisms, the water is not sterile, and within a week of exposure to water, tests confirm glass can develop a biofilm, although it may be thin. So all surfaces in contact with water will have some biological activity on them, even in the presence of low levels of disinfecting agent (that is the levels up to no more than about 0.4 to 0.5mg L^{-1} of available chlorine) that will occur usually in drinking water.

It is clear that while a water satisfies the requirements of the prescribed values in the Water Supply (Water Quality), Regulations, this will not necessarily prevent deleterious action on materials in contact with it, although in practice the control of pH of drinking water means that the higher end of the pH range is achieved, and phosphate dosing up to the maximum may be carried out to limit corrosion of metals, particularly lead.

Developments in CEN in Relation to Effects of Materials on Drinking Water Quality

M. FIELDING

Water Research Centre, Medmenham, UK

1. BACKGROUND

The preceding paper by Mr A. H. Goodman referred to the UK system for the approval of materials and products to be used in contact with drinking water. In other European countries similar systems exist, but often based on different concepts and relying on different testing methods. This has proved a barrier to trade in that a product approved in one country usually needs to be tested again in another country before it can be marketed there. Consequently, in relation to the establishment of an 'open market' in Europe there is a need to harmonise the national standards that relate to the testing and approval of products/materials with respect to possible effects on water quality. The purpose of this short paper is to give a brief summary of what is happening and where we are at present.

2. CEN

The organisation CEN (Comité Européen de Normalisation) was set up to harmonise national standards in general. CEN is supported by the various national standards organisations (e.g. BSI, DIN, AFNOR) of EC member states and EFTA countries and it has the general backing of the EC.

The development and application of harmonised standards is required to meet the requirements of the CPD (Construction Products Directive 89/106/EEC) and, indirectly, other Directives. The CPD requires that products satisfy the 'Essential Requirements'. In the context of drinking water that means a product must not adversely affect the quality of water with which it is in contact, i.e. it must have satisfied the appropriate standard test methods and approval criteria.

In order to develop harmonised test methods and procedures a myriad of CEN committees, working groups, task groups and *ad hoc* groups have been set in motion. In each supporting CEN country the national standards organisation operates committees that 'shadow' the corresponding CEN committees. When an agreed harmonised standard has been finalised it will be adopted as a national standard in each country by the national standards organisations.

3. DRINKING WATER QUALITY

A network of CEN Working Groups deal with water supply in general and these report to the CEN Committee TC164 (Water Supply). Working Group 3 of TC164, i.e. TC164/WG3, deals with 'Water Quality and Materials'. Reporting to TC164/WG3 are several *ad hoc* groups (AHGs) that have the task of developing and writing harmonised standards in specific areas of interest. Each of these usually has its national 'shadow'! The structure to deal with water quality and materials/ products is as follows:

TC164 'Water Supply'

TC164/WG3 'Water Quality and Materials'(The Netherlands)

TC164/WG3/AHG1 'Organoleptic Assessment'(France)

TC164/WG3/AHG2 'Migration Assessment: Non-metallic products (The Netherlands)

TC164/WG3/AHG3 'Microbiological Assessment'(UK)

TC164/WG3/AHG4 'Positive Lists'(Sweden)

TC164/WG3/AHG5 'Migration from Metallic Products'(Germany).

The countries holding the convenorships are indicated above and the AHGs meet every 3 to 6 months.

4. AHG1 'ORGANOLEPTIC ASSESSMENT'

This group is developing CEN standards for testing products in relation to the imparting of taste (called flavour in CEN), odour, colour and turbidity. It operates as a joint working group with TC155/WG2 (Plastic pipe systems: test methods — water quality) in the area of taste and odour assessment. The test methods being developed as CEN standards are:

1. ORGANIC MATERIALS:

   pipes, fittings, their coatings.
       taste/odour
           - test method
           - interpretation (pass/fail values, etc.)
   colour/turbidity.
           -test methods
           -interpretation.

   tanks, reservoirs, their coatings.
   taste/odour/colour/turbidity
           -test methods
           -interpretation.

Nearly all of the work to date has been devoted to developing a method for the assessment of the potential of organic materials/products to impart taste and odour to water. In the analysis of taste and odour in a quantitative manner it is impossible to define precisely what must be measured since what is measured depends on the method used. So in order to achieve harmonisation the analytical technique for quantifying taste/odour is being 'harmonised' by TC230/WG3 (Analytical methods) and this, when ready, will be built into the test method. Other methods for organic materials will be tackled in due course.

2. INORGANIC MATERIALS:

The testing of inorganic materials, especially metals, for taste and odour has not been considered necessary to date in most countries since control of the appropriate metal ions in water for other reasons has been deemed to render such testing superfluous. However, not all agree with this conclusion and this is an area where 'prenormative research' is proposed to clarify the situation. A complete package of agreed methods for the organoleptic testing of products seems a few years away.

5. AHG2 'MIGRATION ASSESSMENT: NON-METALLIC MATERIALS'

This group is putting together test methods for the assessment of leaching of substances into water from non-metallic materials/products. The methods provide information on how to prepare water samples (that have been in contact with the test product) for subsequent analysis but do not cover what to measure or the level that should not be exceeded in the test. It is not yet clear where this information will be specified. The methods to be produced are more or less as follows:

MIGRATION ASSESSMENT:

    Pipes, fittings, coatings, adhesives
        – test method for factory-made products
        – test method for site-applied products.

    Other products
        – test method for ion-exchange resins
        – test method for membranes.

    Interpretation
        – conversion factors.

The development and harmonisation of methods for factory-made products, such as pipes, fittings etc., is relatively straightforward but since few standards exist in Europe for site-applied materials, such as epoxy resins, cement linings and solvent cements, the development of a harmonised standard for such products is more difficult. The drafting of standards for ion-exchange resins and membranes could be more difficult still and will possibly require 'prenormative' research to validate any proposed methodology.

'Conversion factors'.deal with converting the migration value for a substance leaching from a product, as determined by the test method in the laboratory, to an estimated concentration in the field, e.g. at a consumer's tap. Such a factor (or factors) would need to reflect typical product dimensions and typical, or worst case, operational conditions, especially contact time with water. Work in this area has mainly been advanced by TC155/WG2 (see above) but if such factors are agreed then it is not clear at present how they will be built into harmonised CEN standards.

6. AHG3 'MICROBIOLOGICAL ASSESSMENT'

Testing to see whether products support bacterial growth is important. At present there are two approaches to testing under discussion. Interlaboratory trials are under way on a BSI method and, if acceptable results are obtained, an ENV standard (a draft standard with a two-year duration) will then be produced. However, it will be some time before a final harmonised standard appears.

7. AHG4 'POSITIVE LISTS'

This group is not producing any standards. Its main task has been to provide an inventory of so-called positive lists in use in Europe. 'Positive lists' are lists of approved chemicals (used in materials and products) and corresponding limits on migration. They are used in a few countries in Europe. However, it is important to look into all aspects of approval criteria and not just positive lists, which represent only one potential feature of an approval system. Although a variety of test methods are under consideration in CEN, the associated approval criteria, i.e. the limits for substances that migrate into water from products and the manner in which they are applied to products or materials, are at a very preliminary stage. The implication is that a complete harmonised approval system for products in contact with water is a long way off.

8. AHG5 'MIGRATION FROM METALLIC PRODUCTS'

This group is endeavouring to harmonise test methods or approaches for metallic products. However, migration of metals from such products is markedly dependent on water quality or characteristics, such that a product may be perfectly acceptable in one water quality region but unacceptable in another. Accommodating this behaviour into harmonised European test procedures is proving difficult.

9. SUMMARY

In relation to the 'open market' in Europe, harmonisation of approval procedures for products that are intended to be used in contact with drinking water is of obvious importance. A system for generating these standards has been established but progress is inevitably slow and complete approval procedures (test methods, approval criteria, etc.) will not appear for some years, at least at the present rate of progress. The mere act of trying to standardise different national approaches, which often for good reasons contain arbitrary features, is leading to considerable scrutiny of the underlying justification of traditionally accepted practices. As a consequence, in some areas supportive 'prenormative research' will be needed before reliable methods become available.

There is little doubt that the EC, particularly DG XI, will need to become more involved at some stage, mainly in the area of stipulating or sanctioning the acceptance limits for chemicals migrating from products, where these relate to potential health effects or impinge on the EC Directive on the quality of drinking water.

DISCUSSION

Dr T Hodgkiess of Glasgow University asked whether water contamination by leachates has any implications for PVC pipes and whether this is covered by the regulations. The authors replied that the only concern relates to the leaching of lead from lead-stabilised PVC pipes. Current Water Act Standards limit the maximum concentration of lead to $50 \mu g \, L^{-1}$ and this should pose no problem with PVC pipes. However, the World Health Organisation standard guidelines will recommend a limit of only $10 \mu g \, L^{-1}$ and the implications of this will have to be considered. UK policy is that there shall be minimal release of lead, as far as it is practicable, into the environment in general including into potable water supplies.

Mr H S Campbell of the University of Surrey commented that the rate of leaching of lead from lead-stabilised PVC pipes is greatest when the pipes are first exposed to water. He asked whether any settling-in period is allowed in the standards and was informed that both short- and long-term contamination levels are being monitored and that medical opinions will be sought when sufficient data are available in the near future.

Materials

Corrosion of Cast Iron in Potable Water Service

G. GEDGE

Ove Arup and Partners, London, UK

ABSTRACT

The use of cast iron in potable water supplies to buildings is now essentially limited to the incoming mains. Within buildings themselves, cast iron is now generally only used for pump and valve bodies. In these services the material performs satisfactorily, not through any inherent resistance to potable waters but rather because of the measures laid down to prevent corrosion.

The water bye-laws require that iron pipes are cement-lined and valve/pump bodies are provided with a protective paint coating, usually bitumen based, that has been approved for use with potable waters.

This chapter discusses the successful application of these materials and explores why this has proved to be the case. It will also identify the problems that can arise if these preventive measures are not taken. These potential problems include contamination of the supply, loss of wall thickness and reduction in mechanical performance.

1. INTRODUCTION

Cast irons have a long history of successful conveyance of water supplies which goes back over 500 years. The first recorded use of these materials for domestic supplies was in 1455,[1] when they were used to supply water to Dillenberg Castle in Germany. The pipes shown in Fig. 1 were installed to supply water to fountains at Versailles in 1664 and are still in use to this day.

During the Victorian era cast irons were the preferred material for conveyance

Fig. 1. Cast iron water pipe — Versailles.

of waters, both for cross country pipelines and local distribution networks. Many of these systems have provided over one hundred years of satisfactory service and only now are some starting to require refurbishment. Many of these systems involved large diameter pipes as can be seen in Fig. 2, which shows the installation of a water supply pipe to Birmingham from a reservoir in Wales.

In the first half of the twentieth century smaller diameter cast iron was also widely used within buildings for a wide range of components and fittings. These included valves, pumps and storage tanks. Today the use of cast irons in water supplies is restricted to cross-country pipelines and supplies to buildings where ground conditions are inappropriate for alternative materials. Within buildings the use is restricted to valve components where the bore exceeds 50mm.

Fig. 2. Installation of Victorian cast iron water main.

This more limited use of the materials is not a reflection of unsatisfactory performance but rather of the ease and speed of installation of more modern materials such as plastics. Where cast irons are still used performance is almost always satisfactory and few corrosion problems, from the internal surface, are recorded. This is not because of any inherent corrosion resistance of the materials but because the mechanisms of corrosion and the methods of prevention are well understood.

2. MATERIALS

The term cast iron encompasses a family of ferrous alloys in which the principal alloying element is carbon in excess of 1.7%. By careful control of other alloying elements (such as silicon and manganese) and heat treatment processes a wide range of versatile alloys can be produced. For potable water services two generic types of material are used: Grey and Ductile cast irons. These names reflect a fundamental difference in microstructure and hence mechanical properties (Figs 3 and 4).

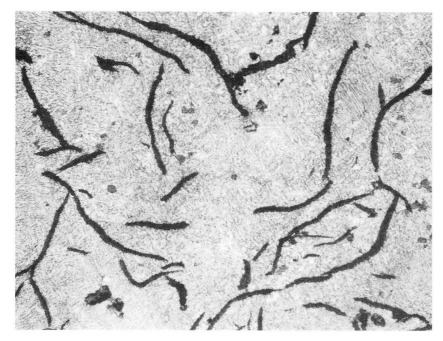

Fig. 3. Grey cast iron — microstructure ($\times$ 300).

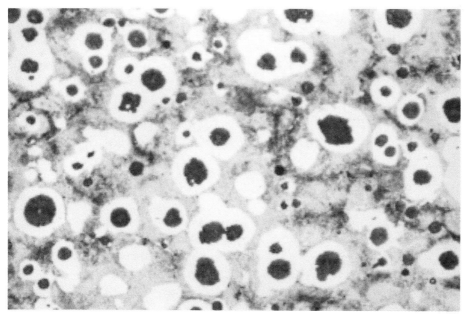

Fig. 4. Ductile cast iron — microstructure ($\times$ 100).

Until the 1950s all material used was of the grey iron type. The microstructure of these materials can range from fully ferritic to fully pearlitic, or mixtures of the two depending on the properties required. However, a common feature which has a major influence of properties is that the carbon exists as graphite flakes. These materials are characterised by reasonably high strengths but poor elongation and impact properties.

In the 1950s it was found that by changing the graphite from flakes to spheroids the elongation and impact resistance could be greatly improved. Because of this improvement the new material was named Ductile Cast Iron (DCI) and this material is now used exclusively for the production of pipes. The use of grey irons is now limited to components of pumps and valves.

3. CORROSION RESISTANCE

Neither DCI nor grey cast irons are usually regarded as being inherently corrosion resistant when exposed to aerated waters. In many respects the materials perform in a similar way to steel components and the basic corrosion mechanisms are essentially the same for both groups of alloys. However, cast irons may give the appearance of greater corrosion resistance for two reasons. Materials used during Victorian times were usually produced with considerable excess wall thickness which effectively becomes a large sacrificial corrosion allowance. Another feature of cast irons that is often mistaken for corrosion resistance is graphitisation. When cast irons corrode it is only the ferrous matrix which is destroyed; this leaves behind the inert graphite. The flakes or nodules can join together and form a skin representing the original component size and shape. This can sometimes lead to the mistaken impression that the material is uncorroded. This is wrong and the layer cannot be relied upon to provide continued integrity.

4. CORROSION MECHANISMS

The corrosion of iron in aqueous media can be explained from a theoretical consideration of the thermodynamic equilibria that occur. These considerations can be displayed graphically in the form of potential–pH diagrams (Fig. 5). These are also known as Pourbaix diagrams after the Belgian chemist who pioneered their development.[4]

In potable water systems the area of interest on the diagram is that around neutrality, pH 6–8, and a potential of approximately −400mV. In this region the diagram indicates that thermodynamics favour corrosion of iron. The diagram also

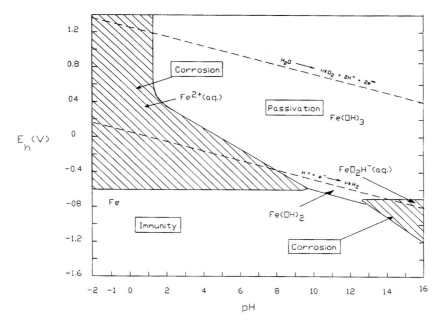

Fig. 5. Potential–pH diagram for iron in water.

indicates that increasing the pH to a value of 12 or making the potential more negative will prevent corrosion.

The mechanism of corrosion can be represented by two simultaneous reactions occurring at different locations on the same surface. At anodic areas iron enters solution generating electrons.

$$Fe \rightarrow Fe^{2+} + 2e^-.$$

This reaction proceeds rapidly in most media, however in aqueous media the reaction rate is usually controlled by the second reaction occurring at the cathode. Here electrons generated at the anode are used in the reduction of oxygen:

$$O_2 + 2H_2O + 4e^- \rightarrow 4OH^-.$$

The overall reaction being summarised as

$$Fe^{2+} + 2OH^- \rightarrow Fe(OH)_2.$$

This ferrous hydroxide acts as a barrier at the iron surface through which oxygen must diffuse for corrosion to continue. Where oxygen availability is low the presence of this layer may retard corrosion. However, in well aerated solutions the hydroxide is rapidly converted to hydrated iron oxide, commonly referred to as rust. This forms away from the surface and is not a protective layer. More detailed descriptions of these mechanisms can be found in standard texts.[2,3]

4.1 Corrosion rates

In practice the use of Pourbaix diagrams is limited as they are based on reversible thermodynamic equilibria. In a corrosion reaction this condition is violated as current flows and the reactions are irreversible. Equilibrium considerations cannot therefore be used to explain corrosion rates. Nonetheless, there is often a remarkably good correlation between predictions from Pourbaix diagrams and real life situations for aqueous media.

As current flows in a corrosion cell either one or both of the reactions may become polarised. In the case of iron the anodic reaction is only slightly polarised whilst the cathodic reaction is highly polarised. The study of polarisation can be used to predict corrosion rates. Polarisation is most usefully expressed graphically as plots of potential vs log current as shown in Fig. 6.

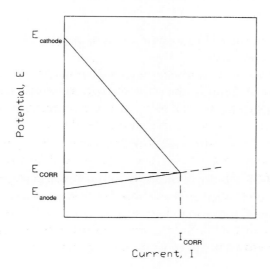

Fig. 6. Polarisation curve for iron in water.

For iron in aerated aqueous media reaction rates are usually controlled by the cathodic reaction and the availability of oxygen. This is because oxygen acts as an efficient depolariser, and the reaction proceeds as rapidly as oxygen can reach the surface.

The effect of increasing oxygen content on the corrosion rate of iron can be seen in Fig. 7. This shows that the corrosion rate initially increases, but beyond a limiting value the rate then decreases. This is attributed to iron passivating because the oxygen is reaching the surface in such excessive amounts that it cannot be used up in the cathodic reduction reaction. Such conditions are unlikely to occur in potable water systems except at abnormally high flow rates.

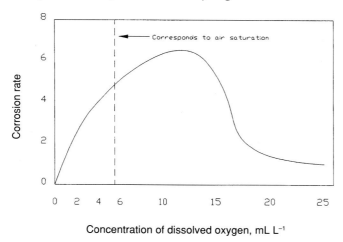

Fig. 7. Concentration of dissolved oxygen (mL L^{-1}).

Oxygen availability is controlled by transport through the solution, by convection or diffusion, and by diffusion across the static boundary layer at the iron water interface. Thus anything that can speed up this transport should result in an increase in corrosion rates. Indeed this is found to be the case.

For potable water systems the most important way in which oxygen availability can be increased is by increasing the flow rate. The effect of flow rate on the rate of corrosion can be seen in Fig. 8. This increase in rate can be explained by the polarisation curves shown in Fig. 9. In this diagram velocity V_1, represents static conditions whilst V_2 and V_3 represent increasing flow rates. The diagram shows that the cathodic curve is depolarised with increasing flow rate and that the current flow, a measure of corrosion, also increases, unless preventive measures are taken.

5. CORROSION PREVENTION

From the preceding discussion it is obvious that iron will freely corrode in potable waters unless preventive measures are taken. The corrosion of iron can cause serious disruption to the operation of water systems. The most obvious problem is general loss of section which can ultimately result in pipe failures. However,

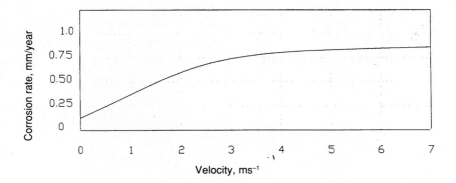

Fig. 8. Effect of water velocity on corrosion.

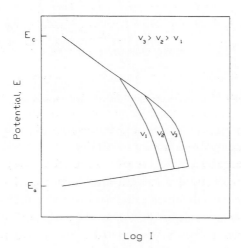

Fig. 9. Effect of velocity on cathodic polarisation curve.

before this stage is reached a number of other difficulties may be encountered.

The products of corrosion tend to be voluminous in nature (Fig. 10), occupying between 5–10 times the original iron volume. Thus the growth of heavy corrosion products on the internal surfaces of pipes will lead to a reduction in bore and a loss of system efficiency. This is further reduced because of the increase in surface roughness leading to an increase in frictional losses.

The corrosion products of iron are not particularly adherent to the surface and can be removed relatively easily. Once in the flow stream, rust may settle out on the surfaces of other components, such as pumps, valves and filters. This can lead to blockages and seizures.

Fig. 10. Corrosion deposits in cast iron pipe.

As far as the water supply is concerned the presence of corrosion products can cause tainting. The iron compounds formed during corrosion are not, of themselves harmful. However they do result in brown discoloration and affect the water's taste. Both these features are undesirable.

The mechanisms and results of corrosion are well known and understood. It has, therefore, been possible to develop control techniques that have proven to be very successful.

In the UK the prevention of internal corrosion is controlled by the water bye-laws.[5] For pipes these regulations require that the internal surface is provided with a cement mortar lining. This is applied in the factory by a centrifugal spray technique. The coating is also provided with a smooth trowelled finish to reduce frictional losses.

In its simplest form a cement mortar lining can be thought of as providing an effective physical barrier between the iron surface and the water. However, these coatings will, to some degree, be water permeable and it might therefore be anticipated that in time water will reach the iron surface and initiate corrosion. This does not happen because of the coatings' inherently high pH at 12–13. It can be predicted from the Pourbaix diagram that the generation of such a high pH at the iron surface moves it into the zone of passivity where corrosion does not occur.

All other components are required to be coated with either bitumen or epoxy based materials. All such coatings have to be approved by the Water Research

Centre for use in potable supplies. Again these coatings provide protection in two ways. Firstly, they act as a physical barrier between the water and iron and, secondly, they impose a large electrical resistance in the corrosion cell. This prevents the flow of ions and electrons between anodes and cathodes, preventing corrosion by highly polarising the reactions.

6. CONCLUSIONS

Cast iron has been used for potable water supplies for in excess of 500 years. Many of the systems constructed during the last century are still operating satisfactorily. The causes and mechanisms of iron corrosion in potable waters are well understood and documented. This understanding has led to the development of preventive measures that have proven to be highly successful.

This long track record, of both the materials and coatings gives a high degree of confidence in their use for potable supplies. This, allied to the knowledge that none of the materials is detrimental to good health means that the internal corrosion of iron components should not be a reason to argue against their continued use.

REFERENCES

1. C. F. Walton and T. J. Opar, *Iron Castings Handbook*. Iron Castings Society Incorporated, 1981.
2. L. L. Shrier, *Corrosion, 2nd Edition*. Butterworths 1976.
3. H. H. Uhlig, *Corrosion and Corrosion Control, 3rd Edition*. John Wiley & Sons, Chichester and New York, 1985.
4. M. Pourbaix, *Atlas of Potential–pH Diagrams*. Pergamon Press, Oxford, 1962.
5. Water supply bye-laws guide, Water Research Centre.

DISCUSSION

Ms K Nielsen of the Engineering Academy of Denmark asked whether the corrosion resistance of ductile cast iron is inferior to that of the earlier grey cast iron because there is less likelihood of a protective skin developing on the corroded surface due to linking up of graphite flakes to form a continuous film. Mr Gedge replied that because of cement mortar lining, internal corrosion is not a practical problem with ductile cast iron pipes and measures can be taken to protect the external surfaces against corrosion. Therefore, no serious corrosion problems are anticipated with ductile cast iron pipes.

Plastics

J. M. MARSHALL

Pipeline Developments Ltd, 2b Skelton House, Manchester Science Park, Manchester, UK

1. INTRODUCTION

1.1 Use of plastics by the water industry

The use of plastics for potable water duties, both in the form of pipes and linings, became widespread in the 1960s. The use of unplasticised polyvinyl chloride (PVCu) pipe began to accelerate in the early 1970s, at which time low density polyethylene (LDPE) was also in use.

Medium density polyethylene (MDPE) was introduced in 1982, and the use of this material soon took over from LDPE for pipe production, increasing markedly from the mid-eighties onwards. Figure 1 shows the percentage use of PE pressure pipe by the UK water industry, and it can be seen that the use showed a steady rise to a level of 24 000 tonnes by 1992. There is little sign of any changes in this trend.

Figure 2 shows the relative usage of the most common materials for water pipes laid in 1991. This depends on diameter, with PE being the most popular for smaller (90–125mm) diameter pipes. Unplasticised PVC takes the lead above this size since it then becomes the cheapest material.

Glass reinforced plastics (GRP) pipes are not often used for potable water supplies, although they are more common for dirty water duties, and they are used only at very large diameters (> 500mm dia.) where iron cannot be used. GRP is used for storage tanks, and often for reservoir applications.

Other polymers, such as polybutylene (PB) and polyvinylidene difluoride (PVDF), chlorinated PVC and cross-linked polyethylene are rarely, if ever used to carry potable water, but are often utilised where temperature resistance is important, e.g. for hot water distribution drainage, central heating, etc., and may thus come into contact with potable water.

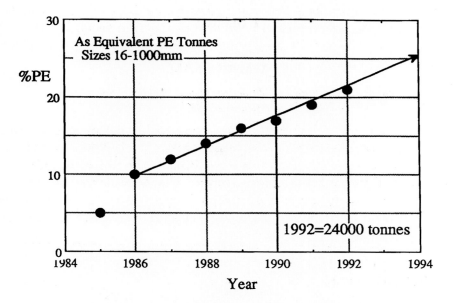

Fig. 1. Percent usage of PE pressure pipe by UK water industry.

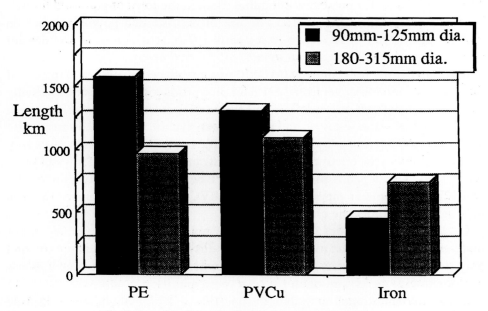

Fig. 2. Lengths of pipe installed by UK water utilities (1991).

There is now a new generation of 'tough' polymers such as high performance polyethylene (HPPE) and modified polyvinyl chloride (mPVC) which have been designed to eliminate problems encountered with the existing materials, and these are beginning to find increasing use in the water supply industry.

The introduction of HPPE now means that the use of PE as a pipe material probably outstrips that of PVC in all sizes.

1.2 CHOICE OF PIPE MATERIAL FOR WATER SUPPLY

Individual water companies have their own policies for pipe material selection, and several have a preference for an all-plastics system. One of the reasons for this is that if the water supply in the area is predominantly from reservoirs, and is 'soft', (often called 'aggressive' because of its carbon dioxide content), then lime can readily be leached from cement mortar lined pipes (all iron pipes are cement mortar lined to protect them from corrosion). This lime causes the pH of the water

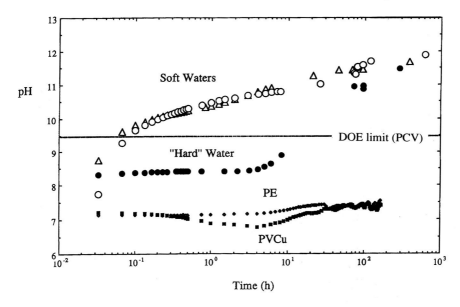

Fig. 3. Increase in pH from cement lining in new ductile iron (cf plastics pipe).

to rise above the level of 9.5 permitted by the DoE regulations — and can indeed elevate the pH to > 11 in extreme cases. Figure 3 shows typical experimental data.

Even areas which have mainly hard water could have problems if the water stands in a cement mortar lined pipe for more than 24h. Plastics pipes — e.g. PVC and PE — do not cause such problems.

The Drinking Water Inspectorate are becoming ever more strict in applying the rules, and now even a short term elevation in pH can give rise to the requirement for an undertaking from the water company.

2. CORROSION OF PLASTICS PIPES

2.1 DEFINITION OF CORROSION

In order to discuss corrosion behaviour, consideration should first be given to what is meant by this.

The definition of corrosion given in the oxford English Dictionary is "to wear away, destroy gradually, decay". Applying this definition to plastics gives rise to the conclusion "Plastics do not corrode".

Indeed the Water Research Centre manuals for MDPE, PVCu and GRP all contain statements that these materials show total resistance to all common forms of corrosion.

This would seem to suggest that plastics are ideal materials and behave perfectly at all times when in contact with water. However, this is not altogether true.

2.2 HYDROLYTIC DEGRADATION

When a material is to be used in contact with water the possibility of hydrolytic degradation must be considered. Susceptible polymers are those containing hydrolysable groups. Polyester resins, which are used for making GRP pipes and storage tanks, fall into this category, as they contain ester $(-\underset{\underset{O}{\parallel}}{C} - O -)$ groups.

2.2.1 Polycarbonate

An application which utilises plastics in contact with water (and also under pressure) is the filter bowl which is used on compressed air lines to remove moisture from the air (see Fig. 4). Polycarbonate has been used for these components, and this polymer contains carbonate linkages within the polymer chain which are susceptible to hydrolysis in a similar manner to the ester groups found in polyester resins.

Polycarbonate filter bowls usually fail in a ductile manner — i.e. by gross yielding and pin-holing (Fig. 4) — but when they have been in contact with water cracks can develop within the wall (Fig. 5) which will ultimately cause brittle failure to occur (Fig. 6).

These internal cracks are caused because, unlike metals, plastics are permeable, and water (and other liquids) can diffuse into the material. Figure 7 (p.35) shows data obtained using a radiotracer technique to follow the ingress of water into polycarbonate when it was totally immersed. Relatively high concentrations of water (0.4–0.5%) were present in the centre of the material after only 48h.

The effect of this absorbed water is then to attack the carbonate–ester links on

Fig. 4. Ductile failure of polycarbonate filter bowl.

the polycarbonate chain, causing chain scission and the evolution of CO_2. This leads to a lowering of the molecular mass of the material, and consequently its mechanical properties, to such a level as to allow cracking to occur. These cracks, even though not always visible, then act as stress concentrators which cause brittle failure of the filter bowl to occur under quite moderate pressures.

2.2.2 Polyester resins
Water will also diffuse into polyester resins and glass reinforced laminates.

A more conventional method of studying water absorption than the radiotracer technique is to measure weight changes when the material is immersed in water and data are shown in Fig. 8 (p.35).

The weight loss exhibited by the sample soaked at 50°C is due to the leaching out of low molecular mass products formed by the hydrolytic degradation of the polyester resin. The result of these reactions is seen as a drop in strength, and this becomes obvious in constant stress tests.

All polymers exhibit creep — i.e. show a decrease in stiffness and modulus with time due to chain rearrangement. Glass reinforced laminates creep less than the resins alone as might be expected, but the creep behaviour of both laminates and

Fig. 5. Internal cracks in polycarbonate filter bowl.

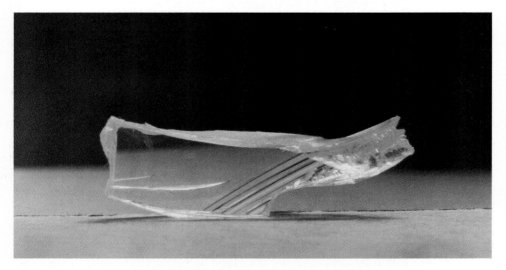

Fig. 6. Brittle failure of polycarbonate filter bowl.

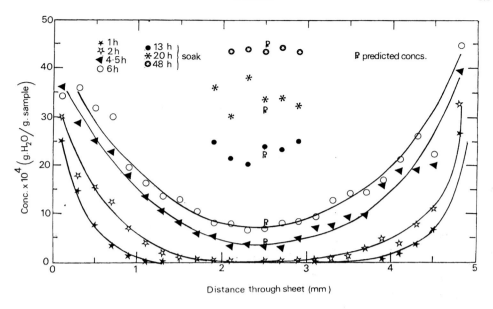

Fig. 7. Diffusion of water in polycarbonate at 80°C — total immersion.

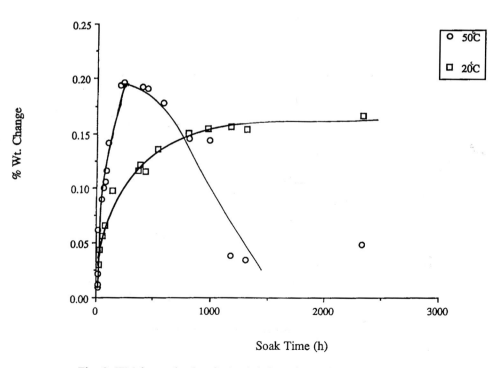

Fig. 8. Weight uptake data for isophthalic polyester laminate in water.

resins is affected by contact with water. The effect of degradation is to cause creep to set in at lower stresses, with materials exhibiting higher creep rates.

However, this is no reason to preclude the use of GRP for potable water duties as potential problems can be avoided by the careful selection of resin type and the use of appropriate test methods.

Resin-based materials (GRP and soft felt resin impregnated liners for sewer renovation use) are tested at constant load in water. The requirement for such testing is included in the relevant Water Industry Specifications (WIS 4-32-02 and 4-34-04) in which minimum physical properties are demanded. The minimum 50 year moduli (extrapolated from long term creep tests) for GRP and soft liners are required to be 5000 and 2200MPa respectively. Typical data obtained from such tests are shown in Fig. 9.

2.2.3 Polyacetal

Another polymer which can suffer brittle cracking when in contact with water is polyacetal, although it is thought that only hot, chlorinated water will affect this material.

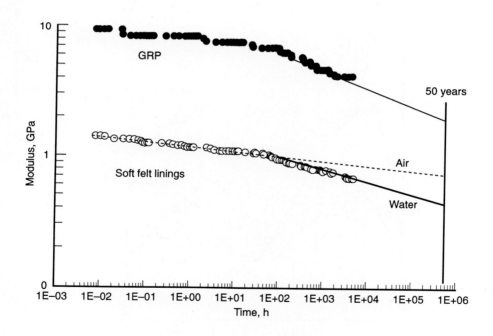

Fig. 9. Effect of water on creep of GRP and soft felt linings.

Plastics

Polyacetal fittings have been used on polybutylene heating systems, both in the UK and more extensively in the USA. Problems have been encountered due to the brittle failure of these components, possibly due to contact with hot chlorinated water, but also to contact with aggressive fluxes containing hydrochloric acid. (Flux is used to clean metal pipes when installing central heating systems — it should not be used for plastics.)

2.2.4 Polyethylene

Plastics pressure pipes are assessed by pressure regression testing and it is well known that PE undergoes a ductile–brittle transition under load at a time dependant on the test temperature. This shows as a change in slope on a log–log graph (e.g. Fig. 10).

Pressure testing is carried out with the pipe immersed in water to effect temperature control, with water also being used for pressurisation. Water will

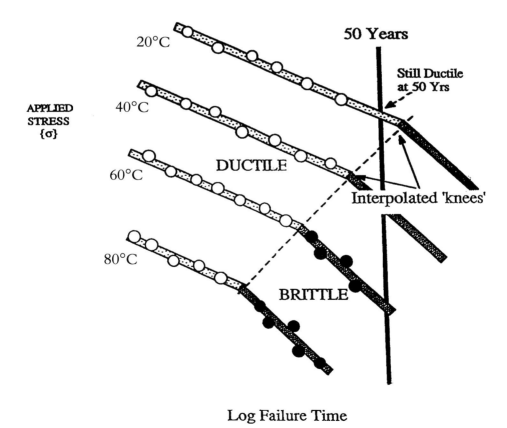

Fig. 10. Ductile–brittle transitions in long-term pressure loading of PE.

diffuse into PE, albeit at very low concentrations (again depending on temperature).

Antioxidants (and other stabilisers) are added to PE to protect it during processing and against any subsequent exposure to heat or ultraviolet light both of which can bring about oxidation of the material. If these stabilisers are removed in any way, or if they are used up, then the material could degrade.

Examples of this have been seen in a poorly stabilised pipe which had been on test in water at 80°C for 400h. The pipe failed because stabilisation was initially poor, and because the stabilisers had been leached out by contact with the hot water. It would not have failed in this time if it had been tested in air at 80°C. This pipe was produced in 1984 and stabiliser packages have been much improved since then. Such occurrences are now extremely rare, and usually due to some error in incorporating the correct stabiliser level into the pipe.

Water Industry Specification 4-32-03 requires polyethylene pipe materials to have an oxidation induction temperature of not less than 230°C when heated in a stream of pure oxygen. Figure 11 shows that exposing a pipe to water at 80°C results in the loss of stabilisers from the pipe wall to a depth of *ca*. 3mm after 4000h.

The practical consequences of this can be seen in Fig. 12, where the rapid drop in pressure carrying capacity observed at long times is due to degradation of the material which is no longer protected by stabilisers.

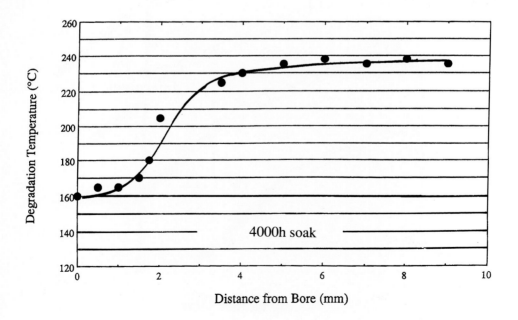

Fig. 11. Degradation in water at 80°C: poorly stabilised PE pipe.

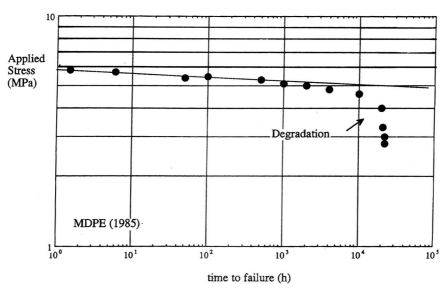

Fig. 12. Reduction in strength of MDPE at 80°C.

2.3 CONTACT WITH SOLVENTS

Solvents will also affect the creep behaviour of PE (see Fig. 13), but they exert only a softening effect, with no degradation of the material. The net result is a loss in pressure bearing capacity, as shown in Fig. 14.

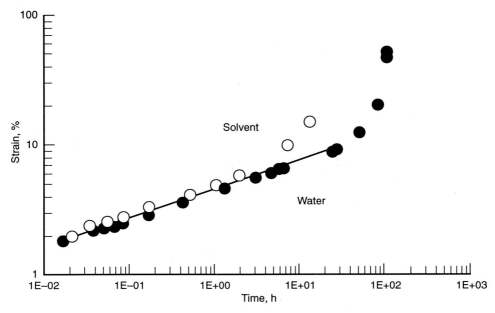

Fig. 13. Creep of HDPE in water and solvent.

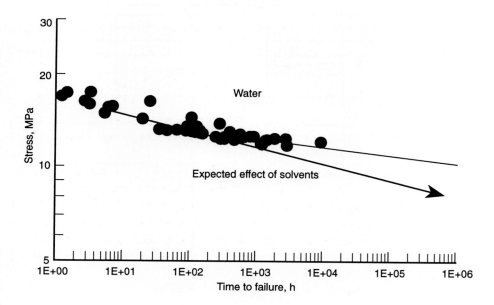

Fig. 14. HDPE regression data.

PVC does not suffer from hydrolytic degradation or environmental stress cracking on contact with water, but can be affected by solvents. As well as being softened by solvent contact, PVC can also be fragmented due to solvent attack. One method of jointing PVC is to use solvent cement (although this is not now approved for water duties), and failures have been seen due to poorly made joints where excess solvent cement has been used.

All the problems discussed above are brought about by extreme conditions, usually involving high temperatures, and field failures can be avoided by the correct choice of materials together with the use of stringent and appropriate test regimes.

Another aspect of pipe material selection which must be considered (although not defined as corrosion) is the effect that any material used to carry potable water might have on that water.

2.4 THE EFFECT OF PIPE MATERIAL ON WATER QUALITY

All polymers for use in contact with potable water must undergo tests to assess their effect on water quality, as set out in WIS 5-01-02. They must comply with Regulation 25 of the water Quality Regulations (1989) and must undergo taste and odour tests and assessment of the degree of extraction of any harmful substances.

2.4.1 Lead extraction from PVC

Currently lead stabilisers are used in the production of PVC pipe, and since lead is a serious source of contamination in drinking water, PVC pipe must comply with stringent lead extraction requirements.

Lead compounds are leached from PVC when it first comes into contact with water. The permitted lead level in water in $50\mu g\ L^{-1}$ (this is likely to be reduced in the near future) and any PVC compound which is to be used to make pipe for potable water duties must get Department of Environment approval for lead extraction levels before it can be used.

Testing is carried out according to BS 6290 section 2.6, and lead levels in the water must drop to the level of detection of the measuring system used ($5\mu g\ L^{-1}$ or less) within a given number of washings. Figure 15 shows the results of lead extraction tests on PVCu pipe produced by UK manufacturers, and it can be seen that the lead levels drop to below the PCV (permitted concentration value) within 2 days in all cases. However, the use of lead stabilisers is to be phased out by 1995 and this will no longer be a problem.

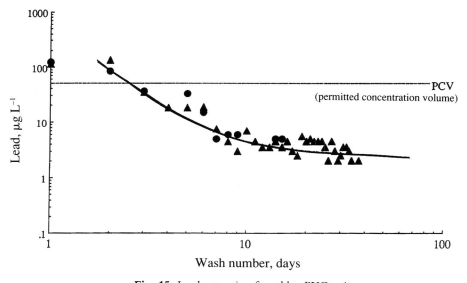

Fig. 15. Lead extraction from blue PVCu pipe.

2.4.2 Permeation of contaminants through pipe walls

Although again not a 'corrosion' problem, ingress of liquids or gases into potable water via the pipe wall is of concern. Water quality can be affected if solvents or other contaminants present in the ground are able to permeate through the pipe wall. All plastics are permeable to some extent, and careful consideration must be given to the choice of pipe materials to be laid in contaminated ground.

(a) PVC

PVC is relatively impermeable, and tends to soften and fail when in contact with any appreciable amount of solvent. The Dutch research institute KIWA have carried out extensive studies in this area, and have produced a number of reports covering the topic.

(b) PE

Polyethylene is rather more permeable than PVC, and studies have been carried out using both solvents such as toluene, chlorobenzene and methanol (industrial contaminants) and disinfectants, herbicides and pesticides (agricultural contaminants).

A radiotracer technique was used and Fig. 16 shows that methanol will penetrate 6mm into a PE pipe wall within 46h at 50°C.

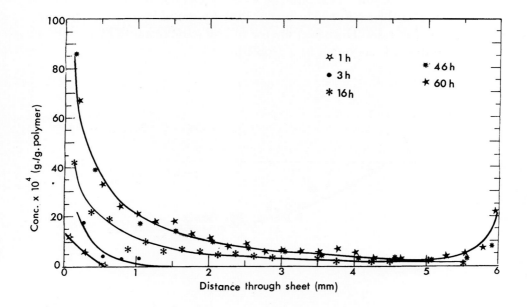

Fig. 16. Diffusion of methanol in HDPE at 50°C.

Chlorobenzene penetrates PE more rapidly and in greater quantities as shown in Fig. 17. However, no trace of the solvent could be found in water surrounding the pipe, even after long exposure times. The release mechanism of the solvent from the pipe wall into the water is obviously of critical importance and this is not yet understood.

It has also been found that the solvent is important in determining the diffusion behaviour of a solute from solution, with permeation into PE usually being faster from an organic solvent than from water.

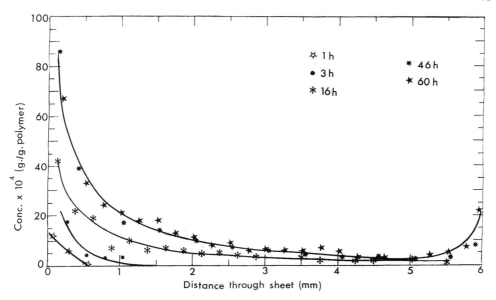

Fig. 17. Diffusion of chlorobenzene in HDPE at 50°C.

Experiments with chemicals having larger, more complex molecules, e.g. herbicides or pesticides, show that these substances penetrate the pipe wall to only a slight extent, if at all, even after long exposure times.

The subject is a complex one, and before laying pipe in contaminated ground, the nature and concentration of any chemicals of concern should be known. Great care must then be taken in the choice of pipe material, but use of plastics may not be entirely ruled out.

3. CONCLUSIONS

1. Plastics do not corrode in the conventional sense when in contact with potable water.

2. Some polymers may suffer degradation (hydrolytic or oxidative), particularly at high temperatures, leading to a loss in strength.

3. This rarely happens under service conditions, and can be minimised or even eliminated by careful material selection and the use of stringent and appropriate tests.

4. Care must be taken when laying plastics pipes in contaminated ground since

some substances may permeate through the pipe wall and adversely affect water quality.

5. Plastics are versatile materials and plastics pipes will give long and trouble free service when properly selected, tested and installed.

DISCUSSION

Dr P Leroy, of CRECEP in Paris, mentioned that calcium salts have been used to stabilise PVC pipes in France for over ten years in order to avoid the contamination of water by lead and that no problems had been encountered. Dr Marshall reported that, in fact, calcium stabilised pipes can have very similar toughness to the lead stabilised materials currently used for potable supplies in the UK.

Mr P F Andrew of Stanton Plc asked whether the studies of lime leaching from cement mortar lined ductile iron pipes in soft water had employed pipes without seal coatings. Dr Marshall agreed that this had been the case but commented that the results are relevant to many mortar-lined pipes still in service which are unsealed.

Dr Hodgkiess of Glasgow University commented that, if properly protected, ductile iron pipes perform satisfactorily, as do grey cast iron pipes. He went on to ask about the relative costs of plastics pipes compared with more traditional materials. Dr Marshall replied that the new modified PVC material should reduce the cost penalty associated with plastics for large diameter pipes.

Concrete: Its Uses in Potable Water Supplies and Interactions with Aqueous Environments

J. FIGG

Ove Arup & Partners, London, UK

ABSTRACT

Concrete is the preferred material for the large-scale containment and distribution of potable water. Large diameter pipes are most economically produced in concrete and a long and trouble-free life can be achieved.

Concrete is about ten times stronger in compression than in tension and therefore reinforcement is necessary to contain bursting stresses. For smaller pipe sizes fibrous reinforcement is effective. Formerly asbestos was widely used but alternative materials must now be considered.

Portland cement-based materials are highly alkaline and therefore ideally suited for the protection of iron and steel from corrosion. However, this same property can result in adverse alkali-aggregate reactions including attack on glass-fibres.

The alkalinity of concrete can cause chemical changes in conveyed water and acidic or low dissolved salts waters can affect the performance of cementitious materials and concretes.

This presentation reviews some of the factors influencing the two-way processes of water–concrete interactions which involve potable waters and considers materials choices available to meet specific problems.

1. INTRODUCTION

Concrete is the preferred material for the large scale confinement and transport of potable water. Pipes of up to 4m in diameter can be made from pre-stressed concrete, as shown in Fig. 1. Man-made lakes, reservoirs, dams and aqueducts are amongst the most dramatic and impressive structures of the world.

Fig. 1. Prestressed concrete cylinder pipe (4m dia.) as used for the Great Man Made River project, Libya.

All such structures depend on the ability of cements to bind aggregates into the artificial conglomerate rock we call concrete. By far the most widely used cementitious binder is Portland cement which is essentially a mixture of calcium silicates (tricalcium silicate and dicalcium silicate).

Portland cement is an hydraulic material. That is it reacts with water to form an hydrated calcium silicate matrix which includes about one-fifth its mass of calcium hydroxide, almost all in the solid crystalline form of the mineral portlandite.

Fig. 2. Concrete gravity dam impounding both surface and desalinated water.

Commercial cements also contain significant amounts of potassium and sodium compounds which during hydration react to form potassium and sodium hydroxides.

Hence Portland cement concretes are highly alkaline so that the pore fluid within the solidified mass has a pH exceeding 13 (approx. 0.7 M with respect to hydroxyl).

Concrete comprises natural or artificial aggregates of graded size (normally maximum 20mm dia.) bound together by the hydrated cement paste to form a strong conglomerate. Unlike stone, however, concrete can be moulded to the desired shape whilst in a plastic state before setting and hardening to form a versatile engineering material.

Concrete is strong in compression: compressive strengths of 60–80 MPa are routinely used and through careful mix design strengths of 100 MPa and more can be achieved. However, the tensile strength of concrete is only about one-tenth of the compressive value and where significant tensile stresses are involved concrete must be reinforced.

2. REINFORCEMENTS

By far the most widely used reinforcement is steel, either conventional mild steel or high tensile steel for pre-stressed concrete. Steel and Portland cement-based concretes ideally complement one another in that they have similar coefficients of thermal expansion, both materials have a high modulus of elasticity and the pH of concrete corresponds with a region of passivation of steel.

Water-containing structures including conveyance pipe are therefore reinforced with mild steel or pre-stressed often by spiral winding with high-tensile steel. All reinforcement must be adequately covered by a layer of high quality concrete to protect the steel from corrosion.

Smaller concrete elements can be fibre-reinforced. Steel fibres are used, but unless stainless steel is employed any fibres at the outside surface will quickly rust leading to an unsightly appearance. Asbestos fibres were widely used in a cementitious matrix for asbestos–cement water pipes and tanks, however, adverse health problems with the manufacture and cutting of asbestos cement militate against continued use (it is believed that swallowing asbestos is not a health problem). Glass fibre reinforcement requires special alkali-resistant glass or alkali-resisting fibre coatings. Polypropylene fibres can confer impact resistance to concretes but the modulus of elasticity of polypropylene is too low for effective reinforcement. Newer polymer fibre reinforcements including polyvinyl alcohol, carbon fibre and aramid are now beginning to come into use and these are capable of true tensile reinforcement.

3. CHEMISTRY AND DEGRADATION

Some siliceous aggregates (and a few carbonate materials) will react with cement alkalies to produce reaction products that can develop swelling pressures within hardened concrete. Such alkali-aggregate reaction can result in a volume increase with cracking and spalling which opens up the structure for further degradation processes (freeze–thaw action, rusting of reinforcement, etc.). An example of such deterioration is shown in Fig. 3. Although rare, adverse alkali reactions must be borne in mind when selecting concreting materials.

Because it is alkaline, concrete is susceptible to attack by acids. Also, like natural stone, concrete is both porous and permeable — the extent depending on the mix ingredients and their proportions and the manufacturing procedure.

Concrete for water-retaining structures should ideally be both impermeable and uncracked, and construction Codes of Practice are intended to encourage production of such material.

Realistically, concrete is always slightly permeable, although if correctly

Fig. 3. Reinforced concrete bollard deteriorated by a combination of freeze-thaw cycling and alkali–silica reaction.

reinforced it should not be unacceptably cracked. Water percolating into hardened concrete can leach sparingly-soluble products, principally calcium hydroxide. The magnitude of the leaching effect depends on the water composition and flow velocity. Very low dissolved salts waters 'hungry waters' and upland acidic waters are the most reactive and attempts have been made to quantify water aggressivity using indexes primarily developed for assessing scale deposition and hence the possibility of attack on steel (e.g. Langelier index, Ryznar Stability index, Driving Force index, Momentary Excess, and Calcium Carbonate Precipitation Capacity).

Like most adaptations, these aggressivity indices are only partially successful and do not give a very satisfactory measure of water/concrete interactions. In part this is due to the very wide range of possible concrete compositions. Computer

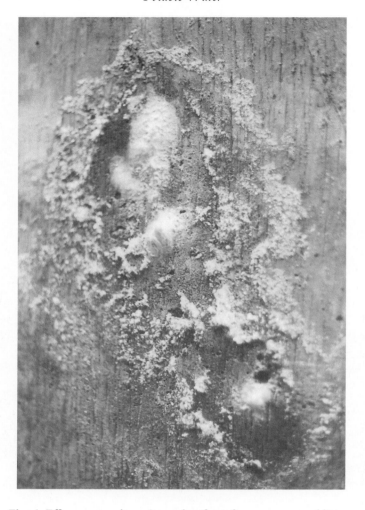

Fig. 4. Efflorescence salts on internal surface of a concrete tunnel lining.

modelling of concrete durability performance in all sorts of circumstances is being attempted by both European and American researchers, the latter under the impetus of the Strategic Highway Research Program. Real waters contain dissolved salts and permeation through concrete to a surface where evaporation can occur can lead to accumulation of efflorescence salts, as illustrated in Fig. 4. Normally this is simply unsightly, but where the actual liquid–vapour transition occurs just below the outer surface, salts accumulation can result in spalling of the concrete and a slow erosion of the hardened material e.g. Fig. 5. On other occasions, the porosity of concrete allows leaching of calcium hydroxide which reacts with carbon dioxide and acid in the air, leading to the formation of stalactites on the surface (as shown in Fig. 6 on p.52).

High sulphate waters can react with the minor proportion of calcium aluminates present in Portland cement concrete. The reaction products result in the generation of internal swelling expansion stresses within the concrete causing volume increase and cracking. A characteristic of such damage is the presence of small acicular crystals of ettringite (tricalcium sulphoaluminate hydrate) in the debris, which in the early days of concrete usage were misidentified as rod-shaped bacteria 'cement bacillus'. Where sulphation reactions occur in the presence of atmospheric carbon dioxide the reaction product is often thaumasite which is structurally similar to ettringite but includes a carbonate grouping.

Acidic sulphates are doubly dangerous and occasionally concrete problems have occurred through incorrect water treatment. An example of this type involved use of potable water in a swimming pool using sodium hypochlorite disinfection. Oxidation with hypochlorite results in hydroxyl ion build up and rising pH of the water which may be lowered by dosing with sulphuric acid. In one municipal pool intermittent batch dosage with H_2SO_4 was employed and, due to information confusion, gross over-additions occurred. This eventually resulted in the dissolution of tonnes of cement grout and mortar (and the generation of petitions to the Medical Officer by parents whose children developed reddened eyes!).

An even more violent reaction occurred in the disinfection contact chambers of one water treatment plant where part-chlorinated water (in practice often over-dosed) was further treated with ozone before pipeline conveyance to the main

Fig. 5. Spalling of concrete caused by evaporation of salt-laden water. This is an example of repetitive action resulting in concentric damage (Liesegang Ring formation).

Fig. 6. 'Straw' stalactites formed by atmospheric carbonation of calcium hydroxide leached by water percolating through concrete.

treatment works. Due to faulty dehumidifiers the silent discharge ozonisers often arced with resultant formation of nitrogen oxides. The 'witches brew' of ozone, nitric oxide and chlorine in effect formed a mixture of hydrochloric and nitric acids, 'aqua regia'. Although severely attacked, the concrete gave a creditable performance compared with steel, aluminium and glass-reinforced polyester which were comprehensively destroyed.

All chemical attack and degradation problems with concrete depend on the presence of water (even damage caused by fire includes steam-pressure rupture and thermal shock cracking due to fire-fighting water jets). Chemical resistance depends on both correct selection of the cementitious system and densification of the concrete mix to minimise permeability.

Since the most vulnerable cement components are the by-product calcium hydroxide and tricalcium aluminates, improved resistance can be achieved by reducing the amounts of these components.

Calcium hydroxide will react with active siliceous materials (pozzolans) such as pulverised fuel ash or silica fume, to form additional calcium silicate hydrate. Latent hydraulic additives such as ground granulated blast-furnace slag if incorporated in concrete are also stimulated to hydrate by hydoxyl ions. Cement formulation can be adjusted during manufacture (albeit at additional fuel cost) to lower the tricalcium aluminate content (one way being to add additional iron oxide to encourage the formation of an alumino ferrite phase in preference to aluminate).

Almost all concrete and concreting problems are due to a lack of understanding of the properties of the material, incorrect or inadequate construction, or misuse of plant or processes. Concrete making is deceptively simple; almost everyone has experience of DIY cement and concrete work. Yet actually concrete technology has advanced tremendously and enormous further improvements can be foreseen.

4. SUMMARY

Concrete has a very long history of successful use in water systems. The Romans were great hydraulic engineers and some Roman aqueducts are still in use today. Roman concretes are lime-based pozzolanic materials but compaction and impermeability are such that carbonation has not progressed to the heart of the material in 1500 years.

Portland cement concrete has a much shorter history. However, some early concretes around 150 years old are still giving satisfactory service. Parts of the wall around Portland Hall (built for William Aspdin, the son of Joseph Aspdin, who patented Portland cement in 1824) at Sheerness have only carbonated to depths of 2–3mm despite outdoor exposure in the Thames Estuary climate.

In fact, concrete is one of our most versatile and durable materials for use in connection with potable water supplies — it just needs understanding.

DISCUSSION

Mr D G Roberts from Fosroc Technology asked whether alkaline leachates from concrete pipes can affect water quality. Mr Figg agreed that this can be so but stressed that it is a two-way process; very low-salt (e.g. moorland) water can be aggressive to concrete by causing the preferential dissolution of calcium carbonate.

Mr N Barraclough of Thames Water asked whether the composition of concrete can be altered to improve its durability. Mr Figg reiterated that the addition of pozzolans such as pulverised fuel ash and silica fume promotes the formation of calcium silicate hydrates which help to fill the spaces between the grains and reduce porosity. The addition of acrylate or styrene polymers as pore blockers can also be very beneficial.

Corrosion of Galvanised Steel in Potable Water Supplies

C.-L. KRUSE

Staatliches-Materialprüfungsamt Nordrhein-Westfalen, Dortmund, Germany

1. INTRODUCTION

In its early stages the corrosion of galvanised steel tubes and vessels in contact with potable water is governed by the corrosion behaviour of zinc. A knowledge of the principles of corrosion of zinc, as well as a knowledge of some special properties of the coating formed on the steel surface during the reaction of the iron with the molten zinc is, therefore, necessary for an understanding of the processes occurring during the attack of galvanised steel.

2. CORROSION OF ZINC

The corrosion of zinc is determined by its electrochemical properties, namely the behaviour of the active metal electrode (anode) and the layer coated oxygen electrode (cathode) and by its chemical and physical properties, namely the solubility and electrical conductivity of the corrosion products.

2.1 General

Zinc in contact with water can be regarded electrochemically as a mixed electrode, on the surface of which at least two electrode reactions occur: the anodic dissolution of zinc

$$Zn \rightarrow Zn^{2+} + 2e^- \qquad (1)$$

and the cathodic reduction of an oxidant. On the basis of its position in the galvanic series, zinc is so active that, in principle, the evolution of hydrogen is possible:

$$2 H_2O + 2e^- \rightarrow H_2 + 2 OH^- \qquad (2)$$

However, in the presence of oxygen

$$^1/_2 O_2 + H_2O + 2e^- \rightarrow 2 OH^- \qquad (3)$$

is the predominant cathodic reaction.

Despite its very active character, in practice zinc reacts relatively slowly. This is due to the formation of a protective layer of corrosion products.

The first protective layer is composed of zinc oxide which is formed according to

$$Zn(OH)_2 \rightarrow ZnO + H_2O \qquad (4)$$

from the primary corrosion product, zinc hydroxide, which itself has practically no protective action. The very thin zinc oxide layer passivates the zinc. As zinc oxide is a semiconductor, the passive layer displays good electrical conductivity which enables zinc to form corrosion cells and undergo pitting corrosion.

In the presence of carbon dioxide another reaction:

$$5 Zn(OH)_2 + 2 CO_2 \rightarrow Zn_5(OH)_6(CO_3)_2 + 2 H_2O \qquad (5)$$

leading to the formation of hydroxycarbonate occurs. This product has a lower solubility than zinc hydroxide (see Fig. 1) and zinc oxide and forms a good protective layer. As this product displays very low electrical conductivity, the development of corrosion cells and, consequently, pitting corrosion is not possible as long as the surface is coated with this corrosion product.

2.2 UNIFORM CORROSION

Using a very simplified model,[1] one can assume that uniform corrosion in cold water is controlled by the growth of the hydroxycarbonate layer to such a thickness that the access of oxygen by diffusion through the layer to the phase boundary of zinc is so restricted that the rate of reactions (1) and (3) becomes equal to the rate of the dissolution of the layer:

$$Zn_5(OH)_6(CO_3)_2 + 8 CO_2 + 2 H_2O \rightarrow 5 Zn^{2+} + 10 HCO_3^- \qquad (6)$$

due to the reaction with carbon dioxide.

This simplified model agrees with the results of investigations[2] showing that the

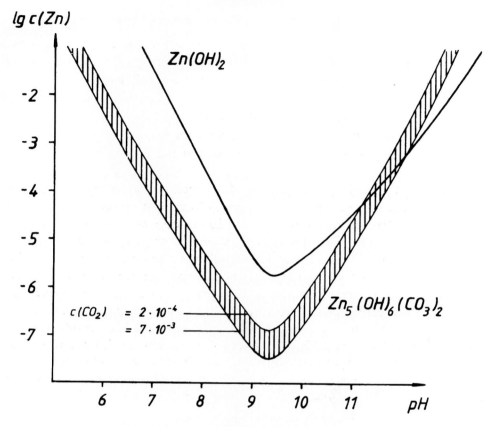

Fig. 1. Solubility of $Zn(OH)_2$ and $Zn_5(OH)_6(CO_3)_2$.

rate of corrosion of zinc in steadily flowing cold water depends only on the concentration of carbon dioxide. As long as the concentration of hydrogen carbonate ions (HCO_3^-) does not vary too much, this fact may be demonstrated by a correlation between the corrosion rate and the pH of the water.

The dependence of the corrosion rate on the flow rate of the water can also be understood with the same simple model.[1] Also, the decrease of corrosion rate with time can be interpreted qualitatively as the result of thickening of the surface layer. A more quantitative consideration of the time dependence of the corrosion rates of zinc leads to a more complicated model.[3]

The influence of the water temperature on the corrosion rate is different from what usually is assumed. Experiments with steadily flowing warm water have shown that the corrosion rate decreases with increasing temperature.[4] This effect, which is associated with the well known ennoblement of electrode potential, is probably a result of the increasing tendency towards the formation of zinc oxide at elevated temperatures.

2.3 Non-uniform corrosion

Non-uniform corrosion is caused by corrosion cells which develop when the reactions of anodic dissolution and cathodic reduction do not occur at the same place. Pre-requisites for the creation of such corrosion cells are:

- variations of electrode potential on the surface;
- surface layers with sufficient electrical conductivity;
- the presence of electrolyte with sufficient conductivity.

Activation of the anodic reaction very often occurs as the result of differential aeration. Under deposits and in crevices with restricted access of oxygen, the electrode potential becomes more negative than in areas with unrestricted access. The stabilisation of the local anodes formed in this way depends on the anion composition of the water. Chloride and sulphate ions which migrate in the electrical field of the corrosion cell into the anodic area form extremely soluble zinc salts. These hydrolyse with water and acidify the anodic area. Zinc hydroxide precipitates as a membrane that separates the anodic area from the bulk solution. The anode is stable as long as more activating anions enter the anodic area by migration then leave by diffusion through the membrane. Inhibition of this process may occur by hydrogen carbonate ions which act as a buffer and remove hydrogen ions formed by hydrolysis. The probability of stable anodes forming can be expressed by the anion concentration quotient Q_{ac}[5]

$$Q_{ac} = \frac{C_{Cl^-} + \frac{1}{2}C_{SO_4^{2-}}}{C_{HCO_3^-}} \qquad (7)$$

with the concentrations, C, in mol L^{-1}. Values below 1 indicate a low probability.

The rate of cathodic oxygen reduction is enhanced by increasing temperature due to the formation of layers of semiconducting zinc oxide. This is accompanied by a shift of the electrode potential towards more noble values, a phenomenon known as potential ennoblement or potential reversal with respect to the electrode potential of iron. In combination with the passivation of the anodic reaction due to the formation of zinc carbonates, potentials that are up to 500mV more noble may be reached. A sudden shift of potential back to the less noble direction may occur after reaching the critical potential for pitting corrosion (pitting potential) which depends on the anion composition of the water and is assumed to become more negative with increasing anion concentration quotient of eqn (7).

Other factors influencing the rate of the cathodic reaction are the concentrations of copper ions and calcium hydrogen carbonate in the water.

Copper ions present in the water may precipitate on the zinc surface according to:

$$Cu^{2+} + Zn \rightarrow Cu + Zn^{2+} \qquad (8)$$

increasing the electrical conductivity of the zinc oxide layer and thereby stimulating the cathodic reaction.

Inhibition of the cathodic reaction may occur due to the reaction of calcium hydrogen carbonate with the hydroxyl ions formed in the course of the oxygen reduction (eqn 3)

$$Ca(HCO_3)_2 + OH^- \rightarrow CaCO_3 + HCO_3^- + H_2O \qquad (9)$$

leading to the precipitation of calcium carbonate at the most active cathodic spots. Calcium carbonate has a very low electrical conductivity and thereby decreases the cathodic area.

3. TYPES OF CORROSION DAMAGE WITH GALVANISED STEEL

The difference in the corrosion behaviour of zinc and galvanised steel results from the fact that the coating does not consist of pure zinc but contains zinc–iron alloy phases formed during the reaction of iron with the molten zinc (Fig. 2). The iron content of the coating increases in the direction of the basis metal.

The probability of corrosion damage in potable water supplies has to be discussed with respect to

- tube quality
- water quality, and
- service conditions.

3.1 Tuberculation

The principle in the protection of steel for use in cold potable water supplies by galvanising is to form a protective layer of iron oxide hydrate phases after consumption of the zinc in the course of normal corrosion. The iron corrosion products which first develop in low concentrations within a matrix of zinc corrosion products may stabilise by ageing if the corrosion rate is sufficiently slow. If the corrosion of zinc proceeds too quickly, corrosion damage may result from the formation of corrosion cells leading to tuberculation, and also from contamination of the water by zinc and iron corrosion products. Tuberculation leads to

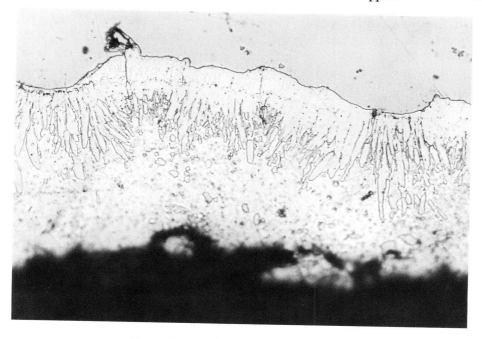

Fig. 2. Cross-section of galvanised steel tube.

a reduction of the effective inner diameter of the tubes and reduces the possible water flow.

A very important factor is the thickness of the zinc coating. The thicker the coating the longer iron corrosion products will be retained in a matrix of zinc compounds. Another aspect of the coating quality is the distribution of the pure zinc and alloy phases. An alloy layer more than approximately half of the coating thickness is believed to be not unfavourable. Areas with a very thin coating may occur around the welding seam of electrically welded tubes if the inner welding seam is not properly machined. In tubes of such quality non-uniform corrosion is very often concentrated around the welding seam.

As long as the corrosion occurs in the pure zinc layer of the coating, the carbon dioxide content of the water will be the only water parameter influencing the corrosion rate. The lower the carbon dioxide content (the higher the pH-value), the lower is the corrosion rate of zinc and the lower is the probability of corrosion damage. As corrosion proceeds and the iron-containing alloy layers begin to corrode, the formation of corrosion cells becomes more likely due to the increasing conductivity of the corrosion products. At this stage, the probability of corrosion damage increases with increasing anion concentration quotient (eqn 7).

The most important influence of the service conditions is the frequency of contact with fresh water. Supply water containing practically no zinc ions dissolves a small amount of corrosion product, losing carbon dioxide and taking up some zinc ions. As a result of this process, the water moves nearer to the point of solubility equilibrium. Under these circumstances the thickness of the layer of corrosion products increases and the corrosion rate is decreased. This is the reason for the well known fact that the corrosion of galvanised steel tubes diminishes with increasing distance from the point of entrance of fresh water.

3.2 Perforation

In the case of warm water, corrosion damage arises mainly in the form of pitting corrosion due to corrosion cells formed because of the electrical conductivity of the zinc oxide layers.

The influence of tube quality arises from the structure of the zinc coating and, in the case of welded tubes, from the properties of the weld seam. When the coating consists totally of zinc–iron alloy phases (which may occur when the products cool down very slowly after galvanising, or when the pure zinc layer is blown off within the tubes with the cleaning steam) the conductivity of the oxide layer is very high from the outset and corrosion cells may form easily. Irregularities in unmachined inner weld seams may initiate anodic behaviour. In such cases, penetration occurs preferentially along the weld seam.

The influence of water quality is the same as with pure zinc.

High anion concentration quotients (eqn 7) increase the tendency for the stabilisation of anodes and lead to an increasing risk of pitting corrosion. Also, the presence of copper ions and the content of calcium hydrogen carbonate are of importance. Water treatment to prevent scaling by calcium carbonate increases the risk of pitting corrosion because the favourable decrease of the cathodic area resulting from calcium carbonate precipitation is prevented.

The main factor influencing the risk of pitting corrosion is the water temperature. Besides favouring the cathodic reaction by forming a semiconducting zinc oxide film, a special type of corrosion forming small anodes from blistering within the coating may occur. At elevated temperatures (above 60°C) the probability of hydrogen evolution according to eqn (2) increases. The hydrogen first forms as hydrogen atoms which may diffuse into the zinc coating. On reaching the alloy layers the hydrogen recombines to molecular hydrogen which is not soluble in the metal lattice and forms blisters. If such a blister cracks and water enters it, the inner part will form a very effective anode of a differential aeration cell.

3.3 Contamination of drinking water

Contamination of drinking water may occur due to:

- soluble zinc corrosion products
- solid zinc corrosion products, and
- rust products.

The amount of soluble zinc compounds in drinking water is practically uninfluenced by the coating quality. It is, however, controlled by the concentration of carbon dioxide and by the duration of stagnation periods. The concentration of zinc ions in water increases linearly with the carbon dioxide concentration. The time dependency is more complicated. After a rapid increase within the first six to eight hours and a maximum after *ca.* 12h a slight decrease may occur due to redeposition of zinc carbonates. In very soft waters, where the predominant corrosion product is zinc hydroxide, higher zinc contents due to the presence of colloidal or suspended corrosion products may be observed.[6]

Solid zinc corrosion products may occur when a certain type of selective corrosion leads to intergranular attack within the pure zinc layer. This type of corrosion is associated with coatings with relatively thick pure zinc phases. The most important factor from the water composition is the concentration of nitrate ions which can act as a passivator. If the concentration is in a critical range passivation takes place only within the grain surface but not at the grain boundaries. The corrosion proceeds into the grains forming a layer of corrosion products with only poor adherence. Turbulent flowing water transports the corrosion products to the tap.[7]

In the later stages of corrosion with rust layers present, water flow after longer stagnation periods or turbulent flow may lead to severe discoloration of the water due to rust products. This type of corrosion damage is favoured by all those factors that increase uniform and non-uniform corrosion of zinc.

3.4 European regulation for drinking water

All requirements for drinking water have to be fulfilled by the water leaving the last tap within a building. The European regulation for drinking water contains two special requirements with respect to the corrosion of copper and galvanised steel. In the case of galvanised steel the allowable zinc content of the drinking water after a stagnation period of 12h is limited to $\leqslant 5\,\mathrm{mg\,L^{-1}}$. As the solubility of the zinc corrosion products depends on the carbon dioxide content of the water, this requirement means that galvanised steel may not be used with waters containing more carbon dioxide than $0.5\,\mathrm{mol\,m^{-3}}$ which is fulfilled with waters with pH $\geqslant 7.3$ and $C_{HCO_3^-} \leqslant 5.0\,\mathrm{mol\,m^{-3}}$.

3.5 Corrosion protection

The possibilities of corrosion protection for galvanised steel may be summarised as follows:

- choice of high quality materials
- water treatment, and
- limitation of water temperature.

The choice of high quality materials consists (in the case of tubes) in the selection of tubes with a sufficiently thick inner coating with an optimally smooth surface and (for welded tubes) a properly machined weld seam. In the standard for galvanised tubes under preparation by ECISS this quality will be characterised as quality A1.

The possibilities of water treatment are limited in the case of drinking water. Treatment within the limits of the drinking water regulations is called water conditioning, and includes neutralisation of carbon dioxide by alkaline or buffering compounds as well as the addition of phosphates and silicates to enable formation of protective layers of corrosion products or lime scale.

A very effective method of stopping corrosion processes within warm water systems is the electrolytic generation of small amounts of aluminium oxide hydrates which to some extent form colloidal solutions. The aluminium oxide hydrate acts as a cathodic inhibitor decreasing the cathodic activity of the surface. Figure 3 shows the results of measurements [8] with coatings consisting of zinc iron alloy phases within different warm water circuits.

The most effective corrosion protection within the field of service conditions is the limitation of water temperature to values as low as possible. The increase of risk of non-uniform corrosion due to increased formation of zinc oxide already begins with temperatures about 35°C.[4] In the region of 60°C the risk in most of the distributed drinking waters is relatively high.

REFERENCES

1. C.-L. Kruse, *Werkstoffe und Korrosion* **26**, 1975, 1, 454-460.
2. C.-L. Kruse, W. Friehe and W. Schwenk, *Werkstoffe und Korrosion* **39**, 1988, 12-23.
3. J. Rückert and D. Stürzenbecher, *Werkstoffe und Korrosion* **39**, 1988, 7-17.
4. C.-L. Kruse, *Werkstoffe und Korrosion* **27**, 1976, 841-846.
5. German Standard DIN 50930 Teil 3, December 1980.
6. E. Meyer, *Schriftenreihe WaBoLu* **52**, 1981, 9-30.

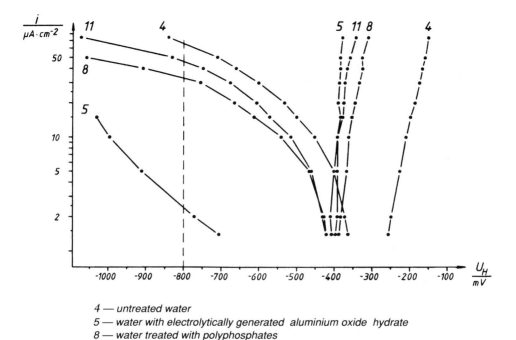

Fig. 3. Polarisation curves for zinc iron alloy phases in different warm water circuits at 60°C.

7. C.-L. Kruse, *Sanitär- und Heizungstechnik* **34**, 1969, 375-377.
8. C.-L. Kruse, Kh.G. Schmitt-Thomas and H. Gräfen, *Werkstoffe und Korrosion* **34**, 1983, 539-546.

DISCUSSION

Prof. W Fischer, of Märkische Fachhochschule, Iserlohn, in Germany, asked whether blisters in the zinc coating on iron are caused by the recombination as gaseous molecules of hydrogen atoms introduced by the corrosion reaction. Dr Kruse agreed that hydrogen atoms liberated by the cathodic corrosion reaction diffuse into the metal and recombine at the iron–zinc interface to generate sufficient hydrogen pressure to cause blistering of the zinc.

Dr Leroy commented that the only way to protect galvanised pipes is by depositing a protective film of calcium carbonate on the zinc coating. Dr Kruse agreed, but only for the case of pitting corrosion in heated water. He stated that in cold water calcium carbonate films do not form on the zinc coating. In a further question, of relevance to water softening, Dr Kruse was asked whether there is any difference from the corrosion standpoint between water containing calcium bicarbonate and water containing sodium bicarbonate. He replied that in the case of uniform corrosion there is no effect of the metal cation; the process is controlled by the carbon dioxide level rather than whether calcium or sodium ions are present. In the case of pitting corrosion softened water is more critical because of the lack of calcium carbonate acting as a cathodic reaction inhibitor.

Copper

J. L. NUTTALL

IMI Yorkshire Copper Tube Ltd, Merseyside, UK

1. INTRODUCTION

Copper is one of the few naturally occurring metals. It has been highly prized through the centuries for its attractive colour, malleability, corrosion resistance and over the last two centuries, for its high conductivity. Archaeological excavations have shown that about 4700 years ago, around 2750 BC, copper pipes were being used for conveying water in Egypt at Abusir in the Nile Delta. Perfectly preserved specimens have also been unearthed at Herculaneum, where life stopped in AD 79 when Vesuvius erupted. The demise of the Roman empire also saw the loss of the artisans demanded by western civilised society where abundant wholesome water is a prerequisite.

Indeed, it was not until this century that copper tubes were used again extensively, replacing the hitherto used lead pipes, which even in Roman times were considered to be dangerous. Around 27 BC Vitruvius, in his book, *De Architectura*, commented: "Therefore, it seems that water should not be brought in lead pipes if we desire to have it wholesome".[1]

Vitruvius would probably have warmly approved of the situation in the UK where more than 95% of tubing used for domestic water services is copper and that its use in other major European countries is growing from their current base penetrations of 40–60%.

Copper is also the dominant material for domestic water systems in North America, Australasia and many countries of the Commonwealth. The third world is now rapidly expanding its use of copper tubing as their societies demand increasing quantities of disease-free water.

The annual world production of copper water tubing is *ca.* 500 000 tonnes, equivalent to 1.25 billion metres or 0.75 million miles. In the UK *ca.* 60 000 tonnes equivalent to 150 million metres or 90 000 miles is installed annually. In global terms UK consumption would encircle the world nearly four times and world consumption over 30 times.

The reason for this enormous consumption of copper plumbing tube is easy to see when one considers its excellent corrosion resistance, its ease of fabrication during installation and hence low installation costs and additionally its contribution to health and the maintenance of wholesome water. Further, copper is seen as environmentally friendly due to its potential to be 100% recycled.

Failures of copper water tube from corrosion are rare but in general well understood events. The high level of quality control exercised in modern tube producing plants plus copper's excellent corrosion resistance gives failure rates that can be measured as less than one in a million. The fabrication of copper during installation is easy due to copper's malleability and the use of soldered and compression fittings which enable sound jointing to be made quickly. A few years ago it was shown that lead in solders did contribute towards lead pick-up in soft waters and as a reaction to that the industry introduced a range of 'potable' integral solder fittings in which no lead is used, the solders being tin–silver or copper–tin.

Copper is classed by some as a heavy metal in close association with lead, antimony, cadmium, mercury, etc. Nothing could give a more erroneous impression of copper. Copper is an essential element for human health.[2] It is well known that we need iron in our diet and many people take iron supplements, yet without copper the iron would be useless. Copper plays a vital part in enzymes and in the uptake of iron into the blood. Animals deficient in copper suffer general ill health, stunted growth due to skeletal changes and various internal disorders. Crops grown on copper deficient soils give rise to stunted growth, poor yield, blemishes and wilting. Fortunately, copper is in many of our foodstuffs but the WHO recommended intake of 2.0–3.0mg per day of copper[3] is not always achieved in Western diets which are estimated at 1.0–2.5mg per day. Fortunately, all normal healthy humans are able to maintain a balance of copper in the body, the main storage area being the liver.

Whilst copper is vital for our metabolism and maintenance of health it is also known to act as a bactericide and fungicide which again contributes to our general health. Work at the Midwest Research Institute[4] has shown that some water borne opportunistic pathogens are killed in copper tube but not in other tubing materials and Schofield and the Public Health Laboratory Service[5,35] have shown that copper has a toxic effect on *Legionella*.

2. WHAT IS POTABLE WATER?

From the dictionary definition it is water in a drinkable form, (from the Latin *poto*, to drink). However, the EC has provided a more detailed definition of water fit for drinking in its directive, EC Directive Relating to the Quality of Water Intended for Human Consumption (80/778/EC);[6] 64 parameters are listed together with Guide Levels and/or Maximum Admissible Concentrations. These cover organoleptic parameters, physico–chemical parameters, undesirable substances in excessive amounts, toxic substances, microbiological parameters and requirements for water softening.

In the UK this directive has been brought into law through the Water Act 1989 and The Water Supply (Water Quality Amendment) Regulations 1989. The standards laid down for water are based on the EC Directive plus 14 national standards. Of the new ideas and concepts brought into force perhaps one of the more radical is that the water should be sampled at the consumer's tap and tested against the prescribed parameters. This specifically includes microbiological as well as metal parameters and states that the water suppliers must, if necessary, treat the water to eliminate or minimise the pick-up of lead, zinc and copper.

Such directives and laws may set many boundary conditions for potable water and standardise its fitness for consumption and to some extent mitigate against corrosion, however, potable water with its widely variable character still poses a considerable challenge as a medium to promote corrosion and degradation of many materials used for its conveyance.

3. COPPER CORROSION IN POTABLE WATER

As a result of its widespread use in nearly all domestic water systems, copper has of necessity come up against most adverse service conditions. In fact, in domestic water installations copper systems can be said to have billions of metre-years' experience. In gaining such experience copper corrosion failures have occurred and in comparison to the tube satisfactorily in use constitute a rare phenomenon. These corrosion instances have been extensively studied and means to resolve the problem have been obtained even if the detailed mechanisms are not always well understood. In many of the cases system design, operation and maintenance have an effect on the occurrence of corrosion.

3.1 Dissolution of copper

The most simple and straightforward type of attack is that due to soft waters with low pH. These waters generally have a pH less than 6, hence they are acidic. In severe cases the water can have a blue–green hue. The problem encountered here is the gradual dissolution of copper until the walls become paper thin and failure occurs. In certain cases the amount of metal loss is extremely small but can nevertheless cause taste and discoloration problems and if no remedial actions is taken continues. Needless to say these waters are outside the EEC[6] mandatory limits for pH which are 6.5–9.5.

This situation can be moderated by treating the water to bring it within EEC limits either at source or as it enters the premises, for example, by passing it through a dolomitic limestone filter.

3.2 Corrosion–erosion

As the name implies this form of attack is the result of a dual mechanism. The corrosion in this instance is the natural oxidation of copper. Under normal circumstances this is limited and protective in the long term. However, in domestic water systems, which are poorly designed or poorly installed, very high water velocities can be encountered. This high velocity water can erode or strip away the protective oxide film. This process of first corrosion (oxide formation) and then erosion takes place repeatedly eventually leading to very local thinning of the walls of the tube and ultimately perforation.

This type of corrosion is particularly noticeable with soft waters where protective films are generally thin. Some trace constituents of the water such as hydrogen sulphide exacerbate the condition. In design terms these systems should not carry water at velocities greater than $2ms^{-1}$ (dependant on temperature) and sharp changes in direction at high water velocity should be avoided.

Installation may also give rise to this problem especially at poorly made joints where the use of blunt or inadequately maintained tools can result in burrs or other local constrictions of the tube bore around which eddy currents form, giving rise to locally very high water velocities of up to ten times the bulk flow velocity.[7] Good design, installation practice and installation quality control checks will avoid this problem.

3.3 Hot soft water pitting

Hot soft water pitting is another rare form of pitting corrosion of copper tubing which is specific to a set of parameters and readily cured by changing one of them. As the name implies hot soft water pitting is limited to hot water systems carrying soft water whose temperatures is greater than 60°C. Mattsson and Fredriksson[8] showed that the composition of the soft water for this type of attack is specific. The water had to be at pH 7.2 or less and the bicarbonate to sulphate ratio of the water had to be less than one. These factors combined with high temperatures can lead to the typical and well-recognised hot soft water pitting (also known as Type 2). The pits are numerous, steep-sided with small cross-section and have grey/black caps of a mixture of copper sulphate and copper oxides. The tube bore is generally covered in light-coloured deposits of silica, alumina and/or iron typical of soft water. This is distinct from the protective black layer of cupric oxide normally formed in hot water in other water areas. Hence hot soft water pitting can be avoided either by reducing the water temperature to 55–60°C, by increasing the pH to greater than 7.2 or by increasing the bicarbonate level to give a bicarbonate: sulphate ratio greater than unity. In practice modification of the temperature is the usual solution and is in any case good practice since water issuing from taps at above 60°C is likely to cause severe scalding.

Alternatively, a design change may be advisable to create a closed (oxygen-free) primary, i.e. indirect, system where temperatures in excess of 60°C are necessary. A modified and more aggressive form of attack occurs when manganese is present in the water at levels greater than 0.02mg L^{-1}. Manganese is a natural constituent of soft water and provided levels remain within the EEC limits[6] it poses no problem for copper.

Work in Japan[9] where the water is continuously dosed with chlorine to high levels has reported a depression in the temperature at which this corrosion takes place. However, since high levels of chlorination of soft waters containing organics is undesirable on health grounds due to formation of trihalomethanes, the problems noted in Japan are unlikely to occur in the UK. The levels of chlorine used in UK public supplies are very much lower than in Japan.

3.4 Type 3 Corrosion

A special form of corrosion in soft water has been described by Linder[10] in Sweden. In this case the corrosion took place in cold waters which were very low in dissolved solids. No protective oxides formed on the tube bores and the pitting was seen as a series of craters beneath mounds of black cupric oxide. The problem was solved by treatment of the water to provide 50–70mg L^{-1} of bicarbonate ion derived from calcium carbonate. Hence the aggressivity of this soft water was reduced by treatment.

3.5 Deposit attack

Waters from upland catchments and some river waters tend to have a high suspended solids burden and contain many natural organic substances. These waters are often acidic and low in dissolved solids so they can cause corrosion and deterioration of iron and concrete mains. Corrosion products may then be picked up and carried with the water, thus adding to the solids burden.

Dependent on system design, operation and maintenance these solids may settle out in tanks, cisterns and pipework in the domestic water system. When these deposits settle on copper tube a corrosion mechanism can ensue. Oxygen cannot penetrate the deposits easily, unlike the remainder of the tube which maintains its protective oxide film. As a result, differential aeration occurs and a corrosion cell is set-up causing pitting beneath the deposit. Copper is not as prone to this type of attack as is stainless steel, for example, but where it does occur perforation may take 10 years or more.

This type of corrosion can be avoided by reducing the ingress of solids.

3.6 Microbially Induced Corrosion

While copper is known to possess bactericidal properties[4, 5, 26, 35] the number and variety of normally harmless bacteria that exist in potable waters mean that some bacteria will be unaffected by copper whilst others will be quickly killed. Sulphate reducing bacteria (SRB) can be a particular problem due to their ability to proliferate in anaerobic conditions. In the oil industry they can turn oil wells 'sour' by the production of acidic hydrogen sulphide which in turn causes considerable corrosion damage to steel pipes. Occasionally, these and other[36] bacteria are found beneath deposits in copper tubes carrying drinking water. Historically, such events have been confined to private supplies where little or no disinfection is carried out or in little used pipework or dead-legs. Such corrosion associated with bacteria is typified by a large number of steep-sided pits grouped together. Beneath the pits is a layer of crystalline cuprous oxide and the corrosion site is capped with sulphates, oxides and organic material. The name 'pepper-pot' pitting has been applied to this form of corrosion which is an apt pictorial description. Whilst SRBs cause water to be unpleasant (in respect of taste and odour) they are of no known consequence to health. Attention to system design and operation to eliminate dead legs and long periods of stagnation usually resolves the situation.

3.7 Flux attack

Flux attack is a result of excess flux being used in the preparation of tube joints. Flux is an aggressive agent meant to chemically clean the outside diameter of the copper tube to provide a surface easily wetted by solder. If flux enters the bore in significant quantities it is not readily washed away and the aggressive constituents continue to attack the copper eventually causing failure. The attack is typically underneath or at the periphery of the flux run.

3.8 Carbon film pitting (Type 1)

The most popular type of copper tube used by the UK plumber is half hard bendable tube. This tube is made by cold drawing a copper hollow to almost finished size; at this stage it is hard due to work hardening. To obtain the half hard temper, the tube is annealed to make it soft and then given a final drawing pass. Up until the 1950s the annealing was carried out by heating the tube in air. This caused heavy oxidation which was removed by pickling the tube in sulphuric acid. This process was known as oxidise–anneal and pickle. It was wasteful in yield terms and created environmental problems. As a result there was widespread adoption of the bright-annealing process where the tube was heated in a reducing atmosphere. The product from this furnace did not require acid cleaning since no oxide was formed; 18 months to two years after the adoption of this process complaints were received of internal pitting corrosion, and the failures increased year by year.

The appearance of this type of corrosion was quite different to types previously noted. Pitting occurred in specific areas in the tube, the spacing between active pits being several centimetres. The general appearance of the tube was the normal blue/green oxide/carbonate coating expected from most cold water services. The pits did not confine themselves to any particular part of the tube as in deposit, flux or impingement attack. The corrosion products were quite dissimilar to those in hot soft water pitting. This type of pitting has been described by the British Non-Ferrous Metals Technology Centre[27] and is also known as 'Type 1' after them. The corrosion products or caps are nodular in appearance and often rise chimney-like above the pit or form small groups of these nodular corrosion caps. Severe hemispherical pitting is usually located underneath the larger single nodules.

3.8.1 The cause of Type 1 pitting

Having had such a startling increase in the number of corrosion failures, a search was initiated to find the cause. After much investigation and some erroneous conclusions, the cause was finally attributed to a layer of carbon in the bore of the tube laid down during manufacture.

Campbell[28] had already noted the potential of such a film for causing corrosion. Several mechanisms for Type 1 pitting have been suggested[29–31] but it has not been positively defined, though the postulated mechanisms have their merits.

3.8.2 Removal of carbon

Having decided carbon was the culprit a way had to be devised to remove it.

High velocity air-borne chilled iron grit to BS 2451 was used and found to be successful in conjunction with improved quality control checks. This is still the main method of cleaning tube in the UK. As the grit impacts onto the surface of the bore of the tube it chips off most of the carbon and breaks up the remainder into small particles. The abraded surface must be at least 80% of the bore if it is to be adequately cleaned. The effectiveness of this treatment can be seen from the reduction of incidence of Type 1 failures from the 1970 peak of 680 cases to less than 30 per annum in the early 1990s (those consisting mainly of old tubes).

3.8.3 The water involved with carbon film pitting

Let us now look at the water quality involved in Type 1 pitting. Water which is derived from surface sources, i.e. rivers, reservoirs, etc. tends to be rich in organic constituents and low in inorganics. Conversely, water drawn from artesian wells tends to be organically pure and rich in inorganics, i.e. hardness salts. Type 1 pitting is almost invariably associated with cold water drawn from artesian wells. It was shown by Campbell[32] and workers at IMI Yorkshire[33] that the presence of an organic inhibitor as a constituent of the water was important in preventing corrosion. They confirmed the presence of organic constituents by using a fluorescence technique and convincingly showed that Type 1 pitting was supported only by a water free from organic constituents and was stopped by using a surface derived water rich in organics.

It has been shown[32, 34] that a water which does not support Type 1 pitting can be turned into a carbon film pitting water by passing it over activated charcoal, thus removing the organic constituent and also that pitting can be stopped if the water is changed to one containing the inhibitor.

4. THE SCOTTISH PROBLEM

4.1 Extent of the problem

In the early eighties the copper tube industry was presented with corroded tubing representative of a corrosion problem in Scotland. The examination revealed the major cause to be differential aeration due to extensive deposits. Samples presented a few years later showed distinct evidence of microbially induced corrosion. Two large institutional buildings in operation for approximately 12–15 years were particularly affected. The main corrosion problem was in the recirculating hot water system and was concentrated mainly on the horizontal portions of the pipework. Additionally, corrosion was found in the cold water system where pipework temperatures had been allowed to rise up to 30°C. It has been notable that this type of corrosion has not been reported and was not evident from private dwellings in the vicinity of the institutional buildings. As a result of a preliminary survey of other similar large buildings there was concern that many, if not all such buildings were affected in the same way. However, a detailed examination of tube samples from 47 sites revealed this particular form of corrosion in only seven, five of which were in the Glasgow area. Examination of corroded samples revealed that tubes of all ages, from 50 years to 18 months, were involved. Since the corrosion only became apparent in the 1980s any historical changes to the copper tube specification or manufacture were eliminated as contributory factors. Speculation concentrated on changes in water quality or operating/maintenance procedures.

4.2 Appearance of tube bores

Copper tubes examined from the two severely affected buildings in Glasgow showed significant amounts of peaty organic deposits. Indeed the removal of the 159mm dia. main return line in the domestic hot water system revealed exceedingly thick, heavy gelatinous deposits which raised many questions regarding the water system cleanliness. Examination of the affected tubes showed that there was some superficial corrosion due to deposit attack (differential aeration) but that perforation was at sites containing pepper-pot pits. In cross-section the pits showed a conical grey cap of corrosion products consisting mainly of copper sulphate and cupric oxide which was often hollow. Removing this cap would reveal the pepper-pot cluster of pits.

The morphology of the pits varied slightly with regard to the number and size within a group and whether sulphides could be positively detected. The pits appeared to be in a membrane of cuprous oxide or copper and beneath that were crystals of cuprous oxide extending from the perforations outward in a hemispherical fashion. Further examination of these tube bores has shown that a biofilm can always be lifted from the surface and stained using the periodic acid-Schiff stain.[11] This shows the presence of polysaccharides (by-products of bacteria) which form in copious quantities when biofilms develop.[24]

Tests for biofilms on tubes from other parts of the country have given a 50/50 split between positive and negative biofilm results. Therefore, this special form of corrosion in Scotland is always associated with a biofilm whereas tubes from elsewhere may or may not exhibit a biofilm regardless of their reason for removal.

4.3 Water analysis and water system survey

Three surveys have been carried out by the Water Quality Centre, Thames Water, chosen because of their expertise and experience with bacteria in water systems. Their first survey involved two institutional buildings both of which were affected with this special form of corrosion, one being worse than the other. The survey showed that the water supplied to the buildings was quite typical for the catchment area and not a cause for concern. However, it was noted that the assimilable organic carbon (AOC) level was high when compared with say London water. AOC is a measure of the fraction of the carbon content which can be utilised by micro-organisms. Of more importance was the remarkable drop in dissolved oxygen content of the water, as the system became quiet during the night time period (Fig. 1).

Operational problems were identified. The conclusion drawn from the evidence collected was that the system was fouled with active micro-organisms and evidenced by the drastic drop in dissolved oxygen. Furthermore, steps ought

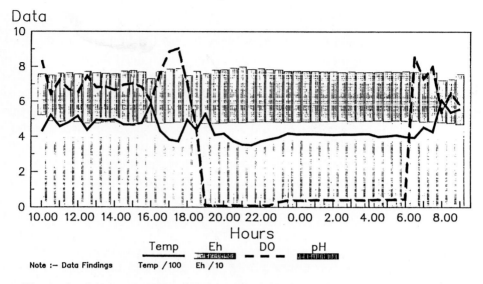

Fig. 1. Thames Water survey results showing dramatic drop in dissolved oxygen over a 24h period.[40]

to be taken to correct the situation and conform to modern standards. The second survey included two buildings known to have this form of corrosion and two which were considered to be free from it, the latter two being outside the Glasgow area. This survey has shown a correlation between those buildings suffering microbially induced corrosion with the measurement of a significant drop in dissolved oxygen overnight, too low a hot water temperature and a high AOC content of the water.

The conclusion drawn was that the water supplied to these buildings was capable of supporting significant microbial growth and that by operating the systems at temperatures known to promote multiplication of bacteria, biofilms form which in turn are capable of initiating and probably sustaining corrosion of the tube by a mechanism yet to be defined. It was recommended that simply operating and maintaining cold water temperature at less that 20°C and hot water temperature greater than 50°C as recommended in the Badenoch Report[12] and the DHSS Code of Practice[13] could largely solve or at least prevent a recurrence of the corrosion problem.

The third survey of an affected building in Glasgow and an unaffected building in Edinburgh showed that where water temperatures were at recommended values bacterial activity had been minimised.

4.4 Corrosion research and theories

This corrosion phenomenon in Scotland is apparently restricted geographically.

It affects large institutional buildings according to their operation and maintenance practices. The copper tube industry has supported several research programmes[36-40] with the aim of understanding the necessary parameters to induce, sustain and control the corrosion as well as determining the mechanisms of corrosion. These research programmes have given the following conclusions:

(a) With regard to the organic (non-microbial) content of water there is no evidence of a direct relationship with corrosion. However, there is a relationship between deposits and pitting initiation but quite unlike deposit attack. Humic materials may play a part in the corrosion mechanism due to their power to chelate and exchange metals.

(b) With regard to the inorganic content of water, further information was added to the already known passivating roles of the anions HCO_3^-, SiO_3^{2-}, HPO_4^{2-} and the aggressive anions Cl^- and SO_4^{2-}. No notable affects on corrosion resulted from excessive pre-chlorination. Study of the Scottish water showed an inhibition of the cathodic process preventing the usual formation of cupric oxide in hot water.

(c) With regard to the microbial content of the water and biofilms, it was shown that only a few copper tolerant bacteria were able to survive with copper but these grew most at temperatures around 45°C, i.e. outside the DHSS recommended levels. Further, raising the temperature to 55°C and above created a synergistic effect between the temperature and copper, rapidly removing the bacteria from the copper surface. This effect could not be matched by the glass control.

It has also been shown quite clearly that certain bacteria can initiate pitting.

With regard to mechanisms, there are two theories neither of which has been conclusively proved.

The first is that bacteria establish themselves under deposits and multiply due to favourable conditions (food source and temperature). One metabolic by-product is hydrogen sulphide which is quickly converted to copper sulphide. The potential difference which exists between copper and copper sulphide is sufficient to drive a corrosion cell. Other bacteria such as *pseudomonads* can create large volumes of polysaccharides which in turn can cause corrosion.[25] Work at Surrey University using identical vessels containing copper tube sections and flowing water, one with and one without bacteria associated with corrosion, has shown that the presence of the bacteria creates corrosion whereas the water alone does not.

A second proposition utilises an existing theory presented by Lucey.[14] Here bacteria induce initial corrosion which is then sustained by a mechanism similar to the Lucey membrane theory, the membrane being largely the biofilm.

4.5 CHEMOSTAT WORK

One line of research which is producing particularly interesting results is that being carried out by the Public Health Laboratory Service, Centre for Applied Microbiology and Research in conjunction with Thames Water using a chemostat.[17] This is a device which realistically models a water system establishing and maintaining bacterial growth on surfaces which can be subsequently examined. This particular work is being carried out using water supplied to one of the Glasgow institutional buildings. The chemostat with this water and the material to be studied is then seeded with bacteria removed from corroding tube samples. Work to date has succeeded in showing that biofilms can be grown on glass, cPVC and copper and that the optimum temperature for most prolific growth, (thickest biofilm) is *ca*. 45°C. A 'pasteurisation' experiment has shown that by raising the water temperature to 60°C the biofilm is effectively removed from copper but not from other materials (Fig. 2). This shows a synergistic effect of copper and hot water and supports the conclusions already drawn by the Water Quality Centre.

4.6 REPLUMBING

One of the affected institutional buildings in Scotland has undergone phased replumbing. It has been recognised by those responsible for the building that

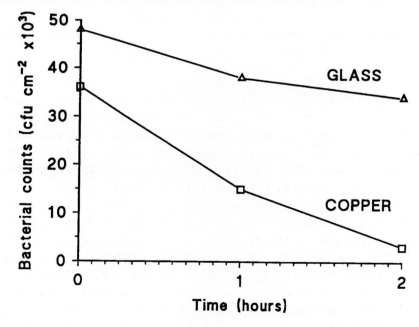

Fig. 2. Pasteurisation at 60°C of a 2 month old biofilm established on copper and glass at 45°C.

significant quantities of suspended matter are not good for the system and as a result, filtration of water down to at least 10μm is being undertaken. As part of an experiment one discrete replumbed section was filtered down to 0.2μm and was also undergoing water treatment to increase the alkalinity of the water. Results were encouraging although the water treatment plant had problems.

4.7. Scottish problem conclusion

From all the research work carried out into this problem to date, it can be concluded that:

(a) Under certain environmental conditions in large institutional buildings copper can be susceptible to microbially influenced corrosion.
(b) A change in these conditions can move a system giving satisfactory performance to one causing corrosion.
(c) This situation should be reversible by cleaning up the system and operating according to DHSS guidelines.
(d) Copper is not particularly prone to such bacterial growth, in fact quite the reverse, hence a change in material without a change in environmental parameters could lead to other microbially influenced problems.

5. *LEGIONELLA PNEUMOPHILA*

5.1. Background

In 1976 in Philadelphia, USA, there was an outbreak of pneumonia of unknown cause among delegates at the Pennsylvania State Convention of the American Legion. A total of 182 cases were recognised of whom 29 died. The disease became known as Legionnaire's 22 disease. The causative agent was eventually shown to be a species of bacterium which was named *Legionella pneumophila* and can cause infection when inhaled as a minute airborne particle. Since then the organism has been studied extensively. It occurs widely in aquatic environments always in association with other bacteria[20] and sources of infection include hot water systems and cooling towers. Although the organisms can be found in many fresh domestic and cooling water systems[18] only certain serotypes are capable of causing disease.

Before an outbreak of the disease can occur certain conditions must be met. Firstly, there must be a site of amplification, the most widely publicised site being cooling towers although domestic hot water systems have been shown to cause more cases.[15] Secondly, a means of transmission, i.e. creation of an aerosol of droplets of less than 5μm dia. which carry the organism, e.g. cooling towers, showers, taps, etc., and thirdly a susceptible population, e.g. those who are ill, immuno-compromised or men over 50 who are smokers. Since the susceptible population will always be with us, as will *Legionella pneumophila*, cooling towers and showers, much effort has been concentrated on minimising or eliminating sites of amplification of the bacterium.

5.2 COPPER AND *LEGIONELLA*

Between 1982 and 1984 the Public Health Laboratory Service carried out a survey of the domestic water services of hospitals, hotels and other large buildings in England and Wales in order to establish the prevalence of *legionellae*.[16] Although the sample size was relatively small, there was some evidence that the incidence of *legionellae* was lower in systems with copper pipework.

This finding was broadly supportive of the work carried out by Schofield[5, 19] who found that when copper was not as well colonised by *legionellae*[4] compared with other plumbing materials. In the USA work carried out on opportunistic pathogens[4] such as *Pseudomonas aeruginosa* showed these bacteria to be susceptible to copper ions. Recent work by the same group has shown the *legionellae* are also inhibited by copper.[22]

The Public Health Laboratory Service in association with Thames Water also carried out chemostat work investigating the effect of copper on *legionellae*.[26] The water used for the experiment was a River Thames derived water which had been artificially softened. This was selected as the best of several waters since it had neither enhanced the growth or death of added *legionellae*. Within the chemostat biofilms of naturally occurring organisms plus *legionellae* were grown on glass slides acting as a control substrate. Following this, copper and polybutylene were used as substrates representing pipework materials. From comparison of the results in Figs 3 and 4 it can be seen that polybutylene allowed the establishment of the thickest biofilm. Copper on the other hand had a suppressing effect on biofilm growth.

The maximum colonisation figures from all the experiments were:

	Maximum Colonisation 10^3 cfu cm^{-2}		Colonisation Ratio	
	Total flora	L. pneumophila	Total flora	L. pneumophila
Copper (aged)	70	0.7	1.0	1.0
Glass	150	1.5	2.1	2.1
Polybutylene	180	2.0	2.6	2.9
Polypropylene	740	66.0	10.6	95.0
Polyethylene	960	23.0	13.7	33.0
uPVC	1070	11.0	15.3	15.7
cPVC	1700	78.5	24.3	112.1

cfu = colony forming units

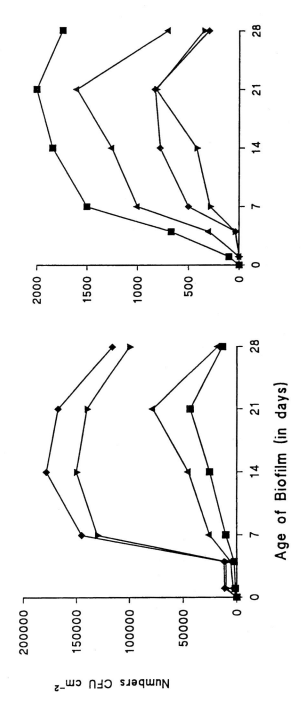

Fig. 3. Comparison of the total flora in the biofilm developing on polybutylene (◆), glass (▼), aged copper (▲) and new copper (■).

Fig. 4. The inclusion of *L. pneumophila* into biofilms on polybutylene (■), glass (▲), aged copper (▼) and new copper (◆).

These results show that compared to glass, copper inhibits the growth of *legionellae* within a biofilm whereas the plastics tested enhance its growth. Further work[40] has shown that polyethylene, polypropylene, uPVC and cPVC all follow the same trend as polybutylene in a variety of waters and temperatures.

The conclusion that may be drawn from the work done by these co-workers is that copper is the preferred plumbing material for controlling the growth of legionellae and further, that incorporation of some types of plastics pipework into an existing copper system may accelerate colonisation by legionellae and other micro-organisms.

6. CONCLUSION

In conclusion, no material is absolutely perfect for every application and this chapter has outlined some conditions where copper is susceptible to failure in potable water. This must be weighed against the generally trouble-free use of 90 000 miles of copper tube installed annually in the UK alone. Over four million miles of copper tubing are estimated to have been installed in the UK during this century and failures are few in comparison. Even in soft waters an exceedingly small proportion fail.

Most of the soft water pitting can be minimised by maintaining EC water standards[6] and if necessary by filtering and increasing alkalinity.

Designers and those who maintain and operate domestic water systems must be careful in their work if they are not to cause problems not only for themselves but also for the ultimate users.[21] Poor design, installation, operation or maintenance (which allows for ingress of extraneous materials, maintaining pipework at temperatures between 20 and 50°C, stagnant or semi-stagnant water, etc.) can lead to ideal conditions for bacterial proliferation which can in turn not only lead to system corrosion but also pose potential health problems.

The studies of microbially influenced corrosion and the establishment of *legionellae* in biofilms on plumbing materials have shown that copper inhibits biofilms and colonisation by legionellae in contrast to other materials some of which appear to enhance microbial growth.

Overall, copper is the only plumbing tube material which is a relatively noble metal with high corrosion resistance also possessing the property to inhibit the proliferation of some bacteria which can cause aesthetic water quality problems and sometimes a potential health problem as with *legionellae*.

7. ACKNOWLEDGEMENTS

The author is indebted to IMI Yorkshire Copper Tube Ltd, for the provision of facilities to enable this paper to be produced; the International Copper Association, IWCC and the UK Copper Tube Manufacturers for funding recent research; and the researchers at the Public Health Laboratory Service, Centre for Applied Microbiology and Research and the Water Quality Centre, Thames Water PLC, National Physical Laboratory, Surrey University and BNF-Fulmer.

REFERENCES

1. L. Vitruvious, Trans. Granger Vol 2 Bk V111, C.6, p. 189.
2. Copper Development Association, Potters Bar, EN6 3AP; Copper in Plant, Animal and Human Nutrition, Technical Note 35.
3. World Health Organisation, Tech Report Series, No. 532, 1973, WHO, Geneva.
4. F. E. Wells, International Copper Research Association Inc., Contractor Report No. 348, Final Report, 1985.
5. G. M. Schofield and R. Locci, *J. Appl. Bacteriology* 1985, 151-162.
6. European Commission Directive Relating to the Quality of Water Intended for Human Consumption (80/778/EC), The Council for European Communities, Council Directive, 15 July 1980.
7. A Guide to Selection of Marine Materials, (A. H. Tuthil, C. M. Schillmoller eds), Ocean Science and Ocean Engineering Conf., Washington DC 1963. Published by International Nickel Ltd., London.
8. E. Mattsson and A. M. Fredricksson, *Brit. Corros. J.* 1968, 3, 246-257.
9. T. Fujii, T. Kadoma and H. Baba, *Corros. Sci.* 1984, 24, 10, pp 901-912.
10. M. Linder and E.-K. Lindman, Proc. 9th Scandinavian Corrosion Congress, 1983, 569-581.
11. A. H. L. Chamberlain, P. Angell and H. S. Campbell, *Brit. Corros. J.* 1988, 23, 3, 197-198.
12. Second Report of the Committee of Inquiry into the outbreak of Legionnaires Disease in Stafford, April 1985. Chairman: Sir John Badenoch, Dec 1987 HMSO, London.
13. DHSS Code of Practice for the Operation and Maintenance of Hot and Cold Water Service Systems, Department of Health and Social Security, 0/11/32145/7. Euston Tower, Euston Road, London. *O11 32 11 457*
14. V. F. Lucey, *Brit. Corros. J.* 1972, 7, 36-41.
15. S. J. Wozniak, Building Research Establishment Information Paper 1983.
16. Public Health Laboratory Service Collaborative Study of *Legionella* Species in Water Systems. HAC November, 1985, 23-27.
17. C. W. Keevil, D. A. Glenister, K. E. Salamon, P. J. Dennis and A. A. West, Int. Biodet. Soc. Proc., 4, 48-62 - Biofilms. (L. H. Morton, A. H. Chamberlain eds).
18. J. S. Colbourne and R. M. Trew, *Israel J. Med. Sci.* 1986, 22, 633-639.
19. G. M. Schofield and A. E. Wright, *J. Gen. Microbiol.* 1984, 130, 1751-1756.
20. R. M. Wadowsky and R. B. Yee, *Appl. Environ. Microbiol.* 1983, 46, 1447-1449.

21. J. L. Nuttall and T. R. Rich, International Tube Association, Conf. Proc., Dusseldorf, Apr. 1988.
22. F. E. Wells, NCRA Contractor's Report No. 348B, 1987.
23. J. Elford, A. N. Phillips, A. G. Thompson and A. G. Shaper, *The Lancet* 18 Feb. 1989, 343-346.
24. J. S. Colbourne and P. J. Dennis, Int. Biodeterioration Soc. Proc. - 7. Biodeterioration. (D. R. Houghton, R. N. Smith and H. O. W. Higgins, eds).
25. G. Geesey, T. Iwaoka and P. R. Griffiths, *J. Colloid and Interfacial Science* 1987, 120, 370-376.
26. J. Rogers, A. A. West, J. V. Lee, P. J. L. Dennis and C. W. Keevil, ICA Project 401, Final Repor, 1990.. ICA, Brosnan House, Darkes Lane, Potters Bar.
27. MP 568, BNF-Fulmer, Denchworth Road, Wantage, Oxon, OX12 9BJ, UK.
28. H. S. Campbell, *J.Inst. Metals* 1950, 77, 345-356.
29. V. F. Lucey, *Brit.Corros. J.* 1967, 2, 175.
30. M. Pourbaix, *Corrosion* 1969, 25, 267.
31. M. Pourbaix, *Corrosion* 1972, 12, 183.
32. H. S. Campbell, *J. Appl .Chem. Cond.* 1954, 63.
33. F. Cornwell, G. Wildsmith and P. Gilbert, *STP* 1976, 576, 159-179.
34. IMI Yorkshire Copper Tube Limited, unpublished work.
35. ICA Project 437, Annual Report 1992, ICA, Lexham House, Hill Avenue, Amersham HP6 5BW, UK.
36. P. Angel, H. S. Campbell and A. H. L. Chamberlain, ICA Report 405, Interim Report, Aug. 1990. ICA, Lexham House, Hill Avenue, Amersham HP6 5BW, UK.
37. R. Francis, R/617/9, November 1990. BNF-Fulmer, Wantage OX12 9BJ, UK.
38. S. Bond, R688/8, March 1992. BNF-Fulmer, Wantage, OX12 9BJ, UK.
39. P. E. Francis and D. E. Meyer, DMM (D) 101, September 1991. National Physical Laboratory, Teddington TW11 0LW, UK.
40. J. T. Walker, C. W. Keevil, P. J. Dennis, J. McEvoy and J. S. Colbourne, ICA Report, 407, Final Report 1988-1990. ICA, Lexham House, Hill Avenue, Amersham HP6 5BW, UK.

DISCUSSION

Mr Davis of R&D Services asked whether the overheating of fluxes during the installation of pipes can lead to deposits on the pipe walls that promote pitting of copper in the same way as carbon films deposited during manufacture can. Dr Nuttall replied that this may be possible and corrosion has occurred by joints in Germany following brazing but not on soldering. Dr Kruse stated that the German problem is not due to flux. He explained that 95% of the corrosion damage in cold water has been due to the formation of copper oxide films generated during high temperature joining at 700°C. It has been demonstrated that these films have the same properties as carbon films in terms increasing the risk of pitting corrosion.

Dr I Wagner of Karlsruhe University disagreed with Dr Nuttall's interpretation of microbial data that copper suppresses biofilm growth compared with glass. He emphasised that results of such experiments not differing by over an order of magnitude must be considered to be identical and went on to assert that differences amounting to large numbers of decades are needed to demonstrate an effect. Dr Nuttall begged to differ, maintaining that the conclusions of this and other work have clearly shown that significant differences exist between the bacteria-forming tendencies of different materials.

Lead: A Source of Contamination of Tap Water

R. GREGORY

Water Research Centre, Swindon, Frankland Road, Swindon SN5 8YF, UK

ABSTRACT

In the UK, lead is absent from virtually all waters entering supply systems. Lead arises in water at consumers' taps because of its presence in various types of pipework and fittings and its uptake involves various mechanisms. The extent to which each mechanism occurs depends mainly on the chemical quality of the water and the origin of the lead, and governs whether the lead is present in the water only as soluble species or also as particulate species. The methods for control of lead in tap water depend on the mechanisms involved in each situation.

INTRODUCTION

Lead has always been a popular metal with 'plumbers' because it generally has a very low corrosion rate and is easy to work with, having, for example, a relatively low melting point. Some Victorians recognised the toxicity of lead arising through its use in water plumbing, but it has only been during the past 30 years or so that the present appreciation of the sub-clinical toxicity of lead has developed. Since the water industry's concern relates to toxicity and therefore with achieving standards, then scientifically the industry is concerned with contamination as distinct from corrosion. The following summarises how lead might come into contact with water, the mechanisms that cause contamination and methods for their control.

2. SOURCES OF LEAD

In the UK, lead is absent from virtually all waters entering supply systems. Lead arises in water at consumers' taps because of its presence in various types of pipework and fittings, including lead run joints, lead pipe, lead-lined tanks, lead solder and brass fittings, and its uptake involves various mechanisms. (Lead is also used as a plasticiser in uPVC pipe and so this can also be a source of lead. However, plastics pipe as a source of lead in tap water is not considered further in this chapter.) The extent to which each mechanism occurs depends mainly on the chemical quality of the water and the origin of the lead, and governs whether the lead is present in the water only as soluble species or also as particulate species.

3. CONTAMINATION FROM LEAD PIPE

The principal source of lead in potable water is lead pipe. Lead pipe was widely used for communication pipes (connecting the main to the stop tap at the curtilage of a consumer's property), supply pipes (from the curtilage to the stop tap just inside the property) and domestic plumbing. The principal mechanism involved is the dissolution of the product of simple oxidation of the lead metal, namely lead carbonate.[1,2] The solubility of lead carbonate is governed mainly by pH, alkalinity and temperature. The effect of temperature is important and is often overlooked. Results from laboratory measurements in Fig. 1 show that increase in temperature increases the equilibrium solubility of lead carbonate. In addition, the rate at which lead dissolves increases with water temperature, so that a higher equilibrium concentration is approached more rapidly in warmer water (Fig. 2). The lead concentrations found in water drawn from a consumer's tap depend not only on water quality but also on the geometry of the plumbing, the plumbing materials used and flow rate. Thus, lead concentrations are greatest when water is warmest, alkalinity is lowest, pH is lowest and the length of lead pipe involved is longest with small pipe diameter and low flow rate. The situation is exacerbated where taps are indirectly fed by lead pipe via a roof tank that might also be lead-lined.

Either of two lead carbonate species can be formed; normal ($PbCO_3$) and basic ($Pb_3(CO_3)_2(OH)_2$). For soft waters with low pH it does not matter which species predominates, because the solubility of both can give rise to unacceptable lead concentrations in tap water,[1-3] as illustrated in Fig. 3 (p.87). However, for hard waters it does matter which carbonate species predominate (Fig. 4, p.87), because only the more soluble carbonate species gives rise to (what are currently considered) unacceptable concentrations. It is convenient to summarise lead solubility data in the form of contours of constant lead solubility in a pH–alkalinity frame as in Figs 5 and 6. Solubility for the least soluble carbonate is shown in Fig.

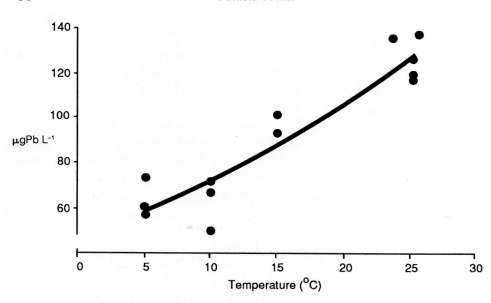

Fig. 1. Effect of water temperature on solubility of lead.

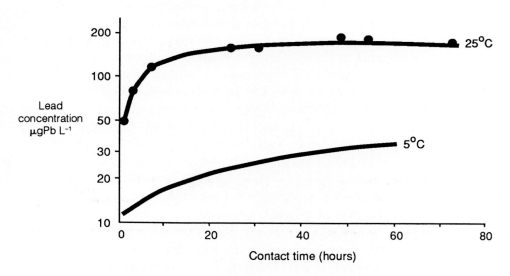

Fig. 2. Rate of solution of lead carbonate at different water temperatures.

5 (p.88). Most importantly, lead solubility decreases rapidly with increase of pH in a low alkalinity water; at pH 8.0 to 8.5 lead solubility is lowest. This prediction for the less soluble lead carbonate suggests that lead should not be a problem in high

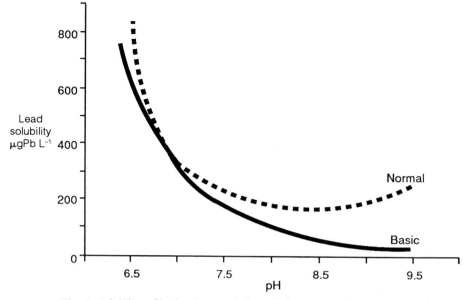

Fig. 3. Solubility of lead carbonates in low alkalinity water (10mg $CaCO_3$ L^{-1}).

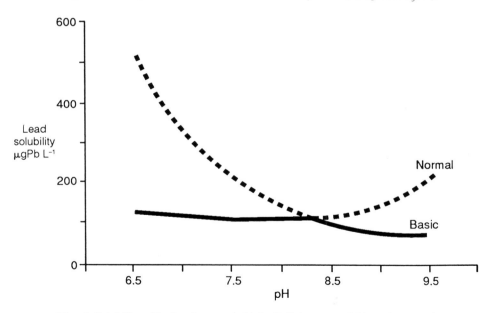

Fig. 4. Solubility of lead carbonates in high alkalinity water (250mg $CaCO_3$ L^{-1}).

alkalinity waters. However, lead is a problem in some high alkalinity waters and this is because the pipe deposits contain basic and not normal lead carbonate.

Figure 6 (p.89) shows the lead solubility contours for basic lead carbonate. The

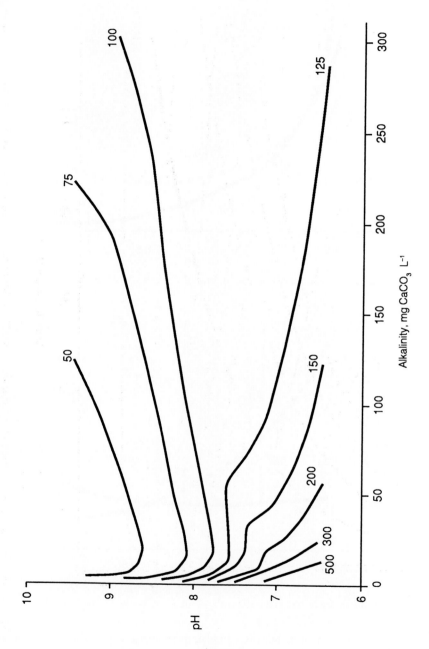

Fig. 5. Lead solubility contours of constant concentration, μgPb L^{-1}, for least soluble corrosion product.

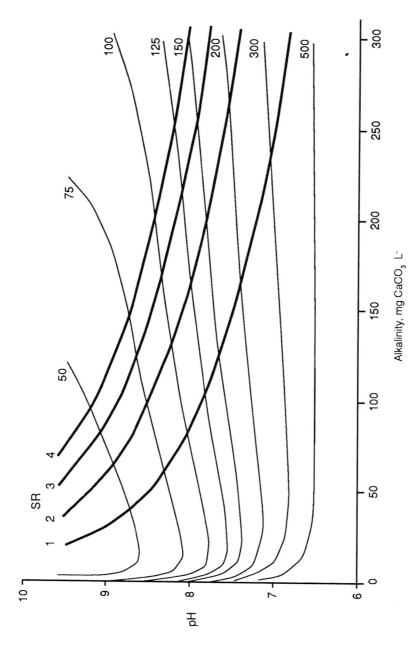

Fig. 6. Lead solubility contours of constant concentration, μgPb L^{-1}, for basic lead carbonate with contours for constant calcium carbonate saturation ratio, SR, superimposed.

contours for calcium carbonate saturation ratio (SR) 1 to 4 have been superimposed to indicate why simple increase of pH is of little use to reduce lead solubility in hard waters: if the saturation ratio becomes greater than *ca*. 1 to 2 then problems will arise from the precipitation of calcium carbonate. (When SR = 1 then the Langelier Index is 0.) With soft, low alkalinity, waters increasing pH to reduce lead solubility does not lead to the possibility of precipitating calcium carbonate. In contrast with hard, high alkalinity, waters reduction of lead solubility is limited by the risk of precipitating calcium carbonate. The diagram shows that alkalinity of hard waters could be reduced (softening) to allow increase of pH to reduce lead solubility but this is expensive.

In soft waters solubility is minimal at pH 8.0–8.5. Therefore, for soft waters, plumbosolvency is controlled by ensuring pH at consumers' taps is not less than 8.0.[2,4] It has been shown in recent years that the occurrence of the more soluble carbonate species is more widespread than previously thought.[3,5] With hard waters, neither pH control nor alteration of alkalinity is a viable strategy. Unfortunately, at present, the reason why the more soluble lead carbonate species should form in preference to the less soluble species is unknown. An understanding of this phenomenon might enable appropriate preventive measures to be developed.

Lead phosphate is much less soluble than lead carbonate. Therefore, control of plumbosolvency in hard waters is achieved through the dosing of orthophosphoric acid, or one of its sodium salts. Phosphate dosing is also advised to supplement control of pH in soft waters. Colling *et al*.[3] have produced a good example of results from sampling at consumers' taps for a hard water dosed with orthophosphoric acid. Their results show well how phosphate takes time to take full effect following the start of dosing. Their results also show the seasonal variation in lead concentration which is mainly due to change in water temperature. The same pattern of effect is found with soft waters. (Orthophosphate is much more effective than polyphosphate — a traditional 'corrosion inhibitor'.[1])

4. CONTAMINATION FROM LEADED SOLDERS

The next most common usage of lead has been leaded solders in copper pipework. The principal mechanism involved is not the same as for lead pipe. In copper pipework the lead solder is galvanically linked. The resulting galvanic corrosion can be faster than the simple oxidation of lead metal. As a consequence of the differences between the two types of corrosion, the morphologies of the corrosion products are different to the extent that the deposit on exposed leaded solder is readily disrupted by flow of water. Consequently, sampling finds lead concentrations which are variable, non-reproducible in replicate samples and which may be greater than might be predicted by consideration of solubility alone.[6]

The rate of galvanic corrosion is governed by different water quality parameters from those governing solubility. The most important quality parameters identified as aggravating galvanic corrosion of leaded solders are the chloride:sulphate concentration ratio (for which alkalinity can be a surrogate) and nitrate concentration. The rate of corrosion appears to be relatively insensitive to pH over the range of interest for potable waters and therefore the pH control strategy is not of much benefit in relation to leaded solders. In contrast, orthophosphate reduces the rate of corrosion and does so even more when zinc is also dosed.

5. CONTAMINATION FROM BRASS FITTINGS

Some of those who have endeavoured to remove the lead problem by replumbing with brass compression fittings have found that lead concentrations can still be unacceptable. It was this experience that drew attention to the small amount of lead used in brass for fittings. With a new fitting, the release of the lead to the water is likely to be governed by either or both solubility and galvanic mechanisms. Recent information suggests that solubility is the controlling mechanism.[7]

6. PARTICULATE LEAD

The existence of what is referred to as 'particulate lead' has been recognised by some for many decades. The pursuit of control of lead levels has drawn attention to the problem of particulate lead.[4,8] Particulate lead can be considered to exist when lead concentrations are greater than could be due to solubility considerations alone. It is recognised to be caused by various mechanisms that include:

- release of deposits formed by galvanic corrosion of lead or leaded solder;
- release of lead-rich deposits from lead pipe or copper pipe downstream of lead pipe, sometimes referred to as flaking lead (with the appearance of tea leaves) — associated with increase of pH for control of plumbosolvency in areas which historically received low alkalinity and low pH coloured waters;
- in association with cement relining of cast iron mains and low alkalinity waters: the cement particles pick up lead from the lead plumbing and the high pH (greater than 9.5) increases lead solubility;
- in association with iron (or manganese): iron oxides are good absorbants of lead. Iron arising from source and not removed at the treatment works, iron coagulant inadequately removed during treatment, and iron arising from

corrosion in distribution picks up lead from the lead plumbing. Iron rich deposits in mains can scavenge lead from the water for decades before it is eventually disturbed. Thus, where iron is involved, a high lead concentration is likely to be found in association with a high iron concentration but not necessarily vice versa.

The established treatments for controlling lead solubility might not control or reduce the production of particulate lead. Control may require quite different strategies. The preferred strategy depends on each situation.

7. THE FUTURE

Currently, The Water Supply (Water Quality) Regulations 1989 (that embody the relevant EC Directive) prescribe a maximum concentration for lead of $50\mu gPb\ L^{-1}$. Other than in extreme situations, lead solubility can be controlled by appropriate water treatment to achieve this level. Even then, control of exposure to lead through water treatment is seen as an interim measure to the eventual replacement of lead pipe and removal of other sources of lead. However, the WHO have recently stated that they intend to revise their Guideline value for lead very substantially from $50\mu gPb\ L^{-1}$ down to $10\mu gPb\ L^{-1}$. This revision has been based on well-established toxicity information. It can be expected that the EC Directive and, therefore, the UK Regulations will adopt this. The implication will be that removal of sources of lead will have to be substantially accelerated because water treatment will become a less adequate approach to reducing lead levels.

REFERENCES

1. I. Sheiham and P. J. Jackson, *J. IWES*, 1981, 35, 6, 491-515, "The scientific basis for control of lead in drinking water by water treatment".
2. R. Gregory and P. J. Jackson, WRc ER 219-S, May 1983, "Reducing Lead in Drinking Water: pH Adjustment and Orthophosphate Dosing to Reduce Lead Solubility".
3. J. H. Colling *et al.*, *J. IWEM*, 1987, 1, 3, 263-269, "The measurement of plumbosolvency propensity to guide the control of lead in tapwaters".
4. R. A. Breach *et al.*, American Water Works Association, Annual Conference, Philadelphia June 1991, pp. 785-819, "A systematic approach to minimising lead levels at consumers taps".
5. J. H. Colling *et al.*, *J. IWEM*, 1992, 6, 3, 259-268, "Plumbosolvency effects and control in hard waters".
6. R. Gregory, *J. IWEM*, 1990, 4, 2, 112-118, "Galvanic corrosion of lead solder in copper pipework".

7. R. J. Oliphant, Foundation for Water Research, Report FR0338, December 1992, "Effectiveness of water treatments in preventing contamination from a leaded brass".
8. A. D. Hulsmann, *J. IWEM*, 1990, 4, 1, 19–25, "Particulate lead in water supplies".

DISCUSSION

Mr G Pascoe of the Cookson Group observed that lead-free solders are now available. He went on to ask Mr Gregory whether, when referring to lead in brasses, he would also include bronzes and gunmetals in the same category. Mr Gregory confirmed that this was his intention. Ms K Neilsen of the Engineering Academy of Denmark stated that their measurements had revealed a big difference between the pick-up of lead by water from brass and gunmetal. In brass, zinc can corrode preferentially whereas in gunmetal lead behaves anodically.

Mr G Phipps of Reliance Water Controls asked whether lead in brass is a significant problem and Mr Gregory stated that it had been on at least one occasion to his certain knowledge.

Dr Leroy of CRECEP in Paris referred to Mr Gregory's belief that in order to conform with the new WHO guideline on lead of $10\mu gL^{-1}$ it would be necessary to remove lead pipes from water systems and asked whether Mr Gregory knew how much this would cost or how it could be done. He went on to state that although there is only 5–10m of lead pipe per house in Paris, it has been estimated that it would take 20 years and cost in excess of £200 million to remove it all. Mr Gregory replied that although the WHO has set a new guideline for lead, the European Commission and national governments have still to agree how it should be adopted. He believed that lead ought be removed from plumbing systems and that priority should be given to those with the longest lengths of lead pipework.

Mr G G Page, a corrosion consultant from New Zealand, contested the assertion by Mr Gregory that although lead has been used for 2000 years, the current appreciation of its toxicity had only developed over the past thirty years. He believed that the harmful effects of lead had been recognised for at least eighty years and stressed that the most serious effects are encountered with small diameter pipes where the surface area to volume ratio is high.

Stainless Steels for Potable Water Systems

D. DULIEU, B. V. LEE, M. O. LEWUS
AND K. W. TUPHOLME

British Steel Technical, Swinden Laboratories, Moorgate, Rotherham, S60 3AR, UK

ABSTRACT

The standard 18%Cr,8–10%Ni austenitic stainless steels are established in handling potable and related process waters in the food, brewing, chemical and pharmaceutical industries, where freedom from contamination is essential. Although offering many advantages, their wider use in potable water distribution systems has been inhibited hitherto by relative cost and their unfamiliarity to specifiers and installers. The principal materials and relevant standards for stainless steels in water systems are summarised in this chapter, together with a brief account of the manufacture of tubing. The basis of the corrosion resistance of stainless steels is described. Effectively the standard 18%Cr, 8–10%Ni austenite steels are inert in cold potable waters and the principal corrosion problems encountered have been associated with poor joining practices. For hot water systems, particularly in heater units, care in vessel design and fabrication is needed. When dealing with waters containing more than ~200ppm chlorides, a more stress corrosion resistant grade is required. A range of more highly alloyed stainless steels is available for aggressive local water conditions and applications in, for example, treatment and effluent plant.

1. INTRODUCTION

Stainless steels are widely used in the chemical, pharmaceutical and food industries for handling waters of varying purity. They offer, in particular, the ability to contain process waters without contamination. More general use in UK domestic water systems has been limited by relative costs and the fact that they are unfamiliar materials to systems specifiers and installers.

There was considerable interest in stainless steels in the UK around the early 1970s, mainly in municipal and institutional installations handling waters which cause difficulties with other materials. Some problems associated with joining methods and workmanship were identified and addressed at that time. However, there have been very few reports on these systems since installation and, to the authors' knowledge, little investigation of their condition, although it is known that some are still operating satisfactorily.

There is now increased interest in the more widespread use of stainless steels, for example in Japan,[1] given the rising standards of water purity and concern over the long term durability and corrosion behaviour of many alternatives.

In this chapter, as an introduction to the corrosion considerations in stainless steel water systems, the current standards for stainless steel water tubing are summarised and the process routes for the manufacture of tubing outlined. The types of fittings used are summarised.

In common with other materials, the successful performance of stainless steels for both cold and hot water system components, including heaters and calorifiers, relies on correct materials selection, good design and fabrication practices and sensible use. Many reported corrosion 'problems' relate to poor design and workmanship, rather than the intrinsic properties of the material itself. In particular, attention must be paid to the quality and cleanliness of welds.

2. STANDARDS FOR STAINLESS STEEL TUBING

Although certain national standards, notably those for Germany and the United States, allow selection from a range of stainless steel grades, the majority call up austenitic 18%Cr, 8–10%Ni or 17%Cr, 11%Ni, 2–3%Mo steels; grades known generically as Types 304 and 316, respectively. Table 1 lists the current UK and German designations for specific variants of these grades and also gives the designations proposed for inclusion in the forthcoming European standard for stainless steels, EN10088.

In the United Kingdom, thin walled stainless steel tubing for water systems is covered by BS 4127: Pt 2 1972 (1986). This describes welded or seamless tubing in the size range 6–42 mm o.d., in the grade 304S15. The standard is undergoing revision and the size range may be extended, with a Type 316 molybdenum-bearing grade being included. Selected standards for stainless steel water tubing within and outside the UK are given in Table 2, where it will be seen that the austenitic steels are widely used. The German Standard DIN 50930 pt 4 gives

Table 1 Compositions (wt%) and designations of standard austenitic stainless steels within the 304 and 316 families

EN10088-1 Steel No	C max	Cr	Ni	Mo	Near Equivalent Grades in:	
					DIN	BS
1.4307	0.030	17.5-19.5	8.0-10.0	-	-	304S11
1.4301	0.060	17.0-19.0	8.0-11.0	-	1.4301	304S31
1.4306	0.030	18.0-20.0	10.0-12.0	-	1.4306	-
	0.060	17.5-19.0	8.0-11.0	-	-	304S15
1.4404	0.030	16.5-18.5	10.0-13.0	2.0-2.5	1.4404	316S11
1.4401	0.060	16.5-18.5	10.0-13.0	2.0-2.5	1.4401	316S31
1.4432	0.030	16.5-18.5	11.0-12.5	2.5-3.0	-	316S13
1.4436	0.060	16.5-18.5	11.0-14.0	2.5-3.0	1.4436	316S53
1.4435	0.030	17.0-19.0	12.5-15.0	2.5-3.0	1.4435	-

Table 2 Examples of national specifications relevant to water tubes

Country	Relevant Specification	Steels Covered	Special Requirements
United Kingdom	BS4127:1972 (1986) (in process of being updated)	304S15	Revision expected to include 316 grades
Germany	D2463 Pt 1 (supply condition) DIN 50930 Pts 1 & 4, (corrosion in water)	Various, including 17 Cr + Ti, 17 Cr + Nb, 18 Cr, 2 Mo, ferritic steels, 304, 304 Ti, 316, 316 Ti, 316 N, austenitic steels and Duplex 22/5	
Denmark	DS439	316	Also covers other materials for water pipework (eg copper, PVC). Water must contain <300 ppm chloride. Only mechanical joining techniques can be used with stainless steel
United States of America	A651-84 (discontinued in 1987)	Types 409, 430, XM8 (18 Cr + Ti), 430 Ti, 434, ferritic steels, 304, 316 austenitic steels	
Japan	JIS G3448 (1989)	Types 304, 316 austenitic steels	Elution test included.

guidance as to the corrosion risks and materials selection aspects for stainless steels.

BS 4825: 1991 is the general food industry specification for relatively large diameter tube, pipe and fittings. This draws upon the seamless tube properties specification BS 3605: 1991 which embraces a range of applications. Within BS 4825 Pt 1 the user may specify from a range of BS 304 and 316 austenitic steel compositions. Traditionally, some use has been made of the titanium or niobium stabilised grades in process plant water systems. These grades overcame the problem of intergranular corrosion resulting from sensitisation, or loss of matrix chromium by precipitation of carbides, in weld heat affected zones. However, the widespread availability of the standard austenitic steels with low carbon contents (Table 1) has reduced significantly the requirement for the stabilised materials.

3. FITTINGS

BS 4127 tube is designed to be joined by either capillary or compression fittings. Capillary fittings may be soldered, but recently, the use of anaerobic adhesives has been accepted into the WRC Water Fittings and Materials Listing for cold water circuits. Both manipulative and non-manipulative compression fittings may be used. Because of the higher hardness of stainless tubing, compared with copper, it is important that olives of suitable ductility and the correct tightening torques are used to ensure leakproof joints. Stainless steel tubes are compatible with copper and copper alloy fittings, but must not be used with cast iron, steel or galvanised ferrous components because of the risk of galvanic attack of these less noble materials.

The German company Mannesmann has developed an integrated tubing and fitting system. This employs a heavier-walled tube than in BS 4127, to allow use of die-pressed compression joints with an internal polymeric O ring to provide the water seal.

4. TUBE MANUFACTURE

Most tubing is made by longitudinal welding. Coils of stainless steel with standard mill finishes are slit, bent to a tube by passing through forming rolls and continuously welded by the TIG, HF or resistance processes. For the smaller diameters, the weld is autogeneous, being formed by melting of the strip edges, without the addition of any filler metal.

Although ductile and readily manipulative, in general stainless steel tubes are significantly stronger than copper tubes of equivalent size, requiring tooling of adequate load capacity for installation work.

Representative internal surface structures of welded tubes are shown in Figs. 1 and 2. Although the inner surface is basically that of the as-supplied strip, tube bores may additionally be cleaned by specialised acid treatments. For special applications requiring locally smooth surfaces electropolishing may be used. Generally, surfaces retaining mill softened and acid-descaled finishes have a high corrosion resistance. The slight grain boundary etching shown in Fig. 1 is a result of the acid descaling operation and reflects the removal of any parts of the metal surface which have been denuded in chromium as a result of oxidation on annealing.

5. THE BASIS OF THE CORROSION PERFORMANCE OF STAINLESS STEELS

Stainless steels are members of the family of metals which are protected by the formation of a self-repairing passive oxide film. Addition of more than *ca.* 11% chromium allows formation of a chromium-rich oxide layer. The stability of this layer increases with chromium content and the general corrosion resistance of the substrate increases with nickel content. At the 17–18%Cr, 8–10%Ni levels present in most austenitic steels used for water tubing, general corrosion attack under conditions met in potable water systems is extremely unlikely.

Penetration of the oxide film by pitting is possible if there is a local concentration of aggressive ions, chlorides being the most common agent in waters, or by differential oxygen levels and aggressive ions, as in crevice attack. The resistance of stainless steels to pitting and crevice attack increases with chromium content and molybdenum and nitrogen additions are also beneficial. A quantitative indication of the relative resistance to pitting of a composition is given by the Pitting Index, PI:[2]

$$PI = wt\%Cr + 3.3 \times (wt\%Mo) + 20 \times (wt\%N)$$

A similar relationship holds for crevice attack.

Although ferritic steels with 17–18%Cr have similar pitting initiation resistance to the austenitic steels, the nickel content in the austenitic steels increases the resistance of the metal matrix to attack and the austenitic steels have generally better formability and weldability.

The molybdenum additions in the 316 austenitic grades (Table 1) confer a significant improvement in pitting resistance, of benefit where operating conditions in respect of temperature and chloride content are unfavourable.

Stress corrosion cracking (SCC) may initiate from metal surfaces under a tensile

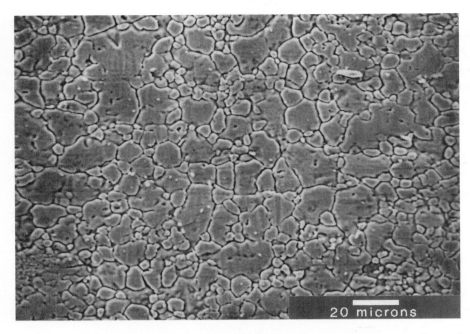

Fig. 1. The internal (bore) surface of a 22mm welded steel tube in the austenitic stainless steel grade 1.4401, showing the slight surface relief caused by the acid descaling process.

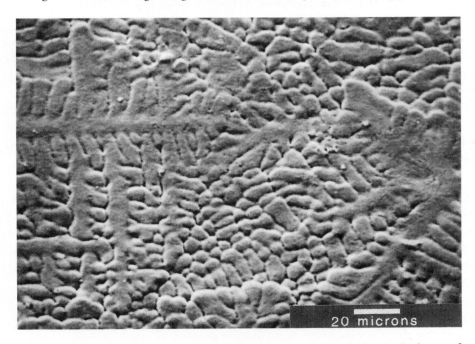

Fig. 2. As Fig. 1, but showing the surface of the weld fusion line in the longitudinal seam of the tube wall (SEM micrographs).

stress, either directly or from corrosion pits. It may occur, usually at temperatures above about 50°C, in aggressive (high chloride, low pH) environments. Generally, the ferritic and duplex stainless steels have greater resistance to SCC than the austenitic steels, although the performance of the latter can be improved by increasing the alloy content, particularly the nickel level.

For normal flow rates, the corrosion performance of stainless steels is insensitive to water velocity and their relatively high strength gives enhanced resistance to erosion in waters containing particulates.

The corrosion resistance of stainless steels is insensitive to pH in the range encountered in potable waters. Deposits may favour the formation of crevice conditions and offer no benefits in protecting stainless steels.

From these basic considerations, it is clear that stainless steels do not rely on the formation of deposits for their corrosion resistance and may be particularly suitable for waters which cause problems with copper.

6. EXPERIENCE IN THE USE OF STAINLESS STEELS — WATER SYSTEMS OPERATING AT NEAR AMBIENT TEMPERATURES

In cold, low chloride potable waters freely exposed austenitic stainless steel surfaces are inert. As an example, McGaul and Geld[3] have reviewed water industry experience and, in their own programme, found little change in the condition of a range of stainless steel coupons, including welded samples, after 13 years exposure in a New York reservoir. The water conditions were generally favourable, (pH 6.8–7.5, 4–7 ppm chloride at temperatures between 2 and 16°C, 6m immersion depth, less than 0.03ms^{-1} flow rate).

However, their review concluded that the standard 300 series stainless steels were suitable, conservatively, for waters with chloride contents of up to 50ppm. German experience[4] indicates that the standard 18%Cr 10%Ni austenitic steels are suitable for potable waters with chloride contents up to 200ppm.

More recently, Japanese experience in the use of stainless steels for piping systems within buildings has been reviewed by the Japan Stainless Steel Association.[1] Tuthill[5] has reviewed the use of stainless steels in both water distribution systems and in treatment plant.

Most of the corrosion problems associated with the use of stainless steels in the storage and distribution of low chloride waters have been attributable to design, workmanship or operational features, rather than to the intrinsic corrosion properties of the standard austenitic grades.

Examples of these problems associated with capillary fittings include very rapid failure of systems as a result of gross contamination from chloride-bearing fluxes and interfacial corrosion, because of the incorrect selection of soldering and brazing alloys.[6] These problems have been addressed in the UK by the issue of a British Standard governing the use of suitable phosphate based fluxes (BS 5245: 1975 (1990)) and the availability of optimised solders.[7] Guidance on suitable joining procedures is available from the principal fittings manufacturers.

Combined crevice and galvanic corrosion has been encountered as a result of leaking flange joints, where the packing medium has contained graphite. This form of attack has long been recognised within the process plant industries and is avoidable by the use of suitable polymeric gaskets or packing.

There is little information on problems of microbially induced corrosion in stainless steel water distribution and communication piping systems. However, the risk of crevice and pitting attack must be considered where conditions of contamination, deposition and nutrient supply allow microbial activity, by analogy with isolated cases reported in water treatment and process plant.[8]

7. STAINLESS STEELS FOR HOT WATER SYSTEMS

As water temperature and chloride levels increase, so do the risks of pitting, crevice and stress corrosion attack of the austenitic stainless steels. These risks are small in potable waters of chloride level around the EEC guideline level of 25ppm.[9] However, they increase at the upper limits for potable waters (for example at the 400ppm maximum value permitted under UK regulations).[10] Guidance on materials performance can be gained from established data sources for stainless steels,[11,12] derived from immersion, electrochemical[13] and SCC testing.[14] Examples from these sources are shown in summary form in Figs 3 and 4. The body of data suggests that the standard 18%Cr–10%Ni steels are satisfactory for waters up to 200ppm chloride at the temperatures encountered in domestic systems. Herbsleb and Tytgat[4] have reviewed the factors affecting materials selection, and give guidance on grades suitable for waters of varying quality, including natural mineral and spa waters.

However, the temperature and circulating water chemistry alone are not a complete guide to corrosion risks, as these can be increased by conditions which increase the severity of the local corrosion environments. As an example, early experience in the UK with use of stainless steels for domestic hot water cylinders demonstrated the importance of design and fabrication factors in the higher temperature regions of water systems. Trials on domestic water heating cylinders in the grade 304S15 gave disappointing results when tested in a high chloride (300ppm) UK water. In part this was caused by weaknesses arising initially from the adoption of cylinder designs established for copper. For example, the crevice created by using an internally domed bottom design provided a site for crevice, pitting and SCC attack (Fig. 5).

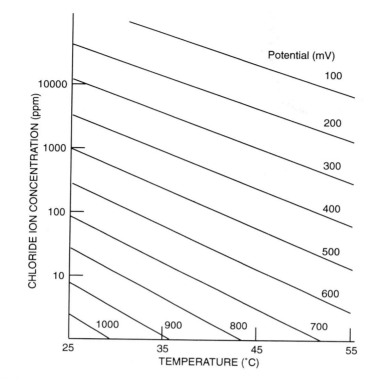

Fig. 3. Pitting equipotential plots against chloride concentration and temperature for the type 304S11 austenitic stainless steel in water of pH 7. (The potentials shown are for a current density of 0.5mA cm^{-2} (Ref.13).)

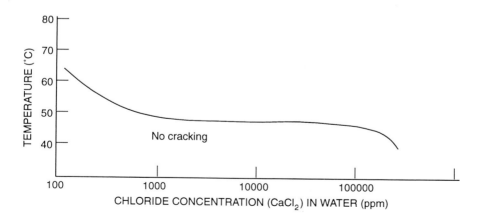

Fig. 4. The temperature-chloride concentration zone for stress corrosion cracking in a static load test to 10 000h duration, 18-10 austenitic steel 1.4306 (Ref. 4).

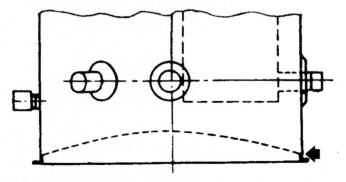

Fig. 5. Pitting and stress corrosion cracking initiated at the crevice created by the resistance seam welded joint at the base of a hot water cylinder made to the traditional design for copper. Material 304S15, wall thickness 0.56mm, base thickness 0.915mm, 12-month duration laboratory test with Clipstone (Mansfield, UK) water of 300ppm chloride.

Magnification ×45.

[Arrow denotes location of seam weld]

Similar experience has been reported elsewhere.[14] Herbsleb [15] has emphasised the importance of good welding and fabrication practices, describing the types of weld defects which may be encountered in fabrication and their implications for corrosion risks.

These studies have emphasised the importance of both good design and fabrication practices. Design improvements include the elimination of crevices at seams and penetrations. For example, the internal bottom dome shown in Fig. 5 may be replaced by an external dome joined to the sidewall by a butt, rather than a seam weld. Welds at potential waterline positions should be avoided. Welding and finishing procedures to give smooth, scale-free inner surfaces are essential.

The importance of surface finishes used in finishing was shown also by Tuthill,[5] who studied the crevice corrosion resistance the Type 304 composition in waters with up to 1000ppm chloride at up to 40°C. This work indicated the deleterious effect of a coarse grit grind over a mill acid descaled finish.

If additional margins of corrosion resistance are needed, for example where operating conditions or system geometry are adverse, more highly alloyed stainless steels are required. German practice indicates the use of the molybdenum bearing grades (Table 1), with their enhanced pitting and crevice corrosion resistance, for chloride levels above 200ppm. The duplex grade 2304 also may be considered. Both austenitic and duplex grades are available with higher pitting and SCC resistance than the 316 range, but these grades are likely to needed for potable water systems only under unusual circumstances, where, for example, mixing of waters may bring about high local chloride concentrations.

In principle, the ferritic stainless steels with higher pitting indices, such as the stabilised 18%Cr–2%Mo grade, offer better resistance to chloride-induced SCC than the standard austenitic grades. However, in common with the higher alloy austenitic and duplex stainless steels, these materials require greater diligence in welding and fabrication.

One problem, to some extent independent of the chloride level in the water, is the risk of attack by the drying-out of leaks or weeps, especially under insulation where relatively high surface temperatures are maintained. Evaporation leads to concentration of salts in the water and can create local conditions far more aggressive than those met within the system.

8. USE OF STERILISING AGENTS

The main action of many sterilising agents used in potable water systems is generation of a highly oxidising solution, to act as a bacteriocide. Agents used include hypochlorites, other halogen compounds, peroxy compounds and ozone.

The basic corrosion risk for stainless steels comes from a combination of chlorides, present naturally or from the decomposition of a sterilising agent, and the highly oxidising condition, which reinforces the cathodic reaction in the corrosion process and leads to enhanced pitting attack. The potential problem is not exclusive to chlorine-containing sterilising additions, and may be encountered if peroxy compounds are used in chloride-bearing waters. Although the main risk is of pitting and crevice attack, the oxidising factor can contribute to SCC under unfavourable combinations of stress and temperature.

Avoidance of the risk of corrosion relies on proper control of the dosing regime. In the food and dairy industries, stainless steel plant is treated frequently with sterilising agents without ill effect, often under automatic control. Risks arise principally from local concentrations of sterilising agents building up, or overlong residence times, as a result of poor dose control or circulation. Breske[16] has reported tests on the effects of contact with halogenated biocides on Types 304 and 316 stainless steels, simulating the inadequate dispersal into systems of solid additions.

The continuous circulation of a low level of chlorine is often required in some industrial and institutional water systems. In principle, at around the 0.1–2ppm chlorine levels used, depending upon the end use of the water, this is a safe procedure for stainless steel systems. Again, with both continuous and 'shock' dosing treatments, the risks of corrosion damage come from poor practices; excessive or uncontrolled dosing levels, local build-up of dosant concentrations within a circuit, and uncontrolled, extended cycle times. In particular, condensation of moist chlorine gas in air spaces within a system can lead to, at least, pitting attack. Where system operations lead to these risks, then consideration should be given to the use of alternative sterilising agents in stainless steel systems.

9. METAL LEACHING

At near ambient temperatures stainless steels may be regarded as inert under most conditions met in potable water systems. Given the concern associated with control over the level of metal ions in potable water, the possibility of transfer of chromium and nickel from a stainless steel surface to water has been examined. Experimental studies have waited until the availability of reliable low level detection and analysis techniques. Measurements have been made by Schwenk,[17] who studied a range of water and surface conditions, and also made observations on samples of tube withdrawn from service in a water system. His work drew attention to the importance of water purity in affecting solvent power and studied the effects of metal surface-to-water volume ratio, the metal surface condition and the time of stagnant contact as factors in determining nickel transfer rates.

The basic mechanism appears to be one of leaching out of nickel from the passive surface film, leaving a more stable, chromium-rich layer. At long stagnation times, nickel appears to be reabsorbed on the pipe surface. The tests on new pipes (X5CrNi1810 austenitic steel), showed that a level of 50µg L^{-1} nickel in the water could be exceeded after standing for one week in new tubes of 15 and 30mm dia. Thereafter, concentrations did not increase further and the nickel migration rate decreased. Water temperature (up to 80°C) and pipe surface condition were not found to be significant factors.

Schwenk concluded that nickel migration occurred only during the initial period after commissioning tubing, until the chromium-rich passive film had been depleted. This was confirmed by tests run on ten year old tubing from a potable water supply system. Nickel concentrations were generally below 7µg L^{-1} and, in most cases, below 2µg L^{-1} in a range of tests at temperatures up to 80°C for durations up to 1680h.

These results are consistent with changes in the stainless steel surface films on exposure[18] and are supported by tests run in the present authors' Laboratories. These have been conducted in accordance with the British Standard Draft for Discussion 201:1991, which gives a standardised procedure for assessing the transfer of metals to water. The test is based on maintaining a fixed surface-to-volume ratio, and exposing a machined and atmospherically aged metal surface to ten batches of an artificial water, of controlled pH and salinity. Water analyses are made on the tenth water sample.

The results obtained for three standard grades of 17–19%Cr stainless steels are shown in Table 3. Included are the acceptance criteria proposed in the Draft and the Water Supply Regulation Guide Level (1989), for comparison. Overall these results support the work of Schwenk in establishing a very low level of nickel and chromium migration into water from the standard austenitic steels.

The initial loss of iron from the surface oxide film is shown in Table 4. X-ray Photoelectron Spectroscopy (XPS) was used to determine the relative metal contents of the oxide film on a surface similar to that shown in Fig. 1.

10. CONCLUSIONS

Stainless steels have many attractive properties for use in potable water systems. They are relatively strong materials, offering the advantages of reduced system weights and resistance to erosion–corrosion at high flow rates. They do not rely on an internal coatings or deposits for corrosion protection and turbulence disrupting deposit formation is an advantage.

At near ambient temperatures in potable waters the standard 18%Cr 8-10%Ni austenitic steels are inert, as indicated by field experience and tests which show

Table 3 Ion guide levels and leaching data for 304L, 316L and 430 stainless steel specimens tested in accordance with BS DD201 (1991)

Sample Type/Guide Levels	Mean Metal Conc μgL^{-1}				
	Cr	Ni	Mo	Fe	Mn
Water Supply Regulation Guide Level (1989)	50	50	50	200	50
BS DD201(1991) Acceptance Criteria (4 x Above Values)	200	200	200	800	200
304S15 (10 Samples) 2 Casts	<5	<5	<5	<10	<5
316S11 (15 Samples) 3 Casts	<5	<5	<5	<10/20*	<5
430S17 (17% Cr Ferritic steel) (10 Samples) 2 Casts	<5	<5	n/a	<10/30*	<5

* Highest Single Value

Table 4 Relative metal contents of oxide film on type 304 2B (hot rolled, softened and descaled) finish, as determined by XPS

	Fe	Ni	Cr
As-received mill finish	0.62	0.05	0.33
After 30 mins in water at 100 °C	0.17	0.14	0.69

levels of metal ion transfer to water negligible in relation to proposed permitted levels.

The steels are tolerant of a range of water chemistry wide in relation to the parameters normally used to define potable waters. In particular, their behaviour is insensitive to pH levels found in potable waters.

The principal factors in assessing corrosion performance in potable waters are chloride concentration and temperature. Below 200ppm chloride a properly designed, fabricated and operated system in the standard Type 304S15 stainless steel is suitable for domestic hot and cold water circulation systems. In hot water heating and storage systems a molybdenum-bearing grade of the 316 type offers enhanced pitting and crevice corrosion resistance. If particular problems of deposit formation, or local chloride level/temperature excursions are expected in a system, then consideration should be given to the use of more highly alloyed compositions. These include the molybdenum alloyed austenitic stainless steels and the duplex stainless steels, which have improved resistance to SCC attack.

However, whatever the grade of stainless steel selected, careful design and workmanship in water systems is essential to ensure optimum performance. In particular welding and surface finishing of welds must be carried out to high standards.

Cases of microbial corrosion of stainless steel systems have been reported, principally in larger scale water treatment and industrial process water plant. The possibility of pitting attack under water conditions allowing microbial activity must be considered.

Sterilisation practices based on oxidising agents liberating chloride ions into a stainless steel system must be used with care. Enhanced pitting, crevice (and possibly SCC) attack may occur as a result of poor, uncontrolled dosing practices.

11. ACKNOWLEDGEMENTS

The authors wish to thank Dr M. J. May, Manager, Swinden Laboratories and Product Technology, and Dr R. Baker, Director of Research British Steel Technical, for permission to publish this paper.

REFERENCES

1. *Piping Manual for Stainless Steel Pipes in Buildings*, Publication No. 12 008, 1987. The Japan Stainless Steel Assn. and the Nickel Development Institute.
2. E. Alfonsson and R. Qvarfort, Proc. Int. Conf. on Electrochemical Methods in Corrosion Research, EMCR '91, July 1991, Helsinki, Finland. Publ. Trans. Tech., Zurich, Switzerland

(also available as the brochure ACOM 1-1992, Avesta Sheffield AB, Sheffield, UK or Avesta, Sweden).
3. C. McCaul and I. Geld, *Materials Performance*, 1978, 5, 27-33.
4. G. Herbsleb and G. Tytgat, *Wasser. Abwasser*, 1988, (129) 3, 153-157.
5. A. H. Tuthill, Conference, Nickel Metallurgy Vol. 2. 11: Industrial Applications of nickel, pp 55-76. Canadian Institute of Mining & Metallurgy, Toronto, 1986.
6. T. Takemoto and I. Okamoto, *Welding Journal*, 1984, 63, (10), 300s-307s.
7. R. C. Randolph and D. A. Lincoln, *Materials Protection & Performance*, 1973 (12), 4, 25-27.
8. S. W. Borenstein P. B. Lindsay, *Materials Performance*, March 1988, pp. 51-54.
9. EEC Council Directive 80/778, Quality of Water intended for Human Consumption.
10. UK Water Supply (Water Quality) Regulations, 1989. Department of the Environment).
11. A. J. Sedriks, *Corrosion of Stainless Steels*, Wiley-Interscience, New York, 1979.
12. Jernkontoret Corrosion Tables for Stainless Steels & Titanium, Stockholm, 1979.
13. J. W. Fielder and D. R. Johns, *Industrial Corrosion*, April/May 1990, 8 (3) 9-14.
14. T. Yoshii *et al.*, in Proceedings 4th Asian Pacific Corrosion Control Conference, Vol. 2, pp.1050-1057.
15. G. Herbsleb, *Mannesmann Forschung*, 1989 (40) 10, 554-559.
16. T. C. Breske, *Materials Performance*, April 1992, 49-55.
17. W. Schwenk, *Brit. Corros. J.*, 1991 (26) 4, 245-249.
18. D. R. Johns, Report EUR 12640EN, The Constitution of Oxide Films and the Corrosion Behaviour of Stainless Steels. The Commission of the European Communities, Luxembourg, 1990.

DISCUSSION

Mr H S Campbell of Surrey University enquired about the use of stainless steels for hot water cylinders. He recalled that many years ago the Stainless Steel Development Association had undertaken field trials but found that failures occurred around the top weld. This was not due to crevice effects but because the weld was oxidised as it was impossible to weld with a protective gas on the inside of the cylinder. Attack of the oxidised weld led to pitting and leakage. He asked whether it was now possible to produce an oxide-free weld on the inside of a closed vessel. Dr Dulieu confirmed that cylinders can be welded with an inert gas inside them but advocated pickling which can be employed with the domestic size of component on an industrial basis. Mr Campbell went on to ask why the BS DD201 test for metal ion leaching had been applied to Type 316L steel when Type 304 steel is much more widely used. Dr Dulieu explained that Type 304 stainless steel is to be evaluated next.*

---

*A Type 304 stainless steel has been evaluated, giving similar performance and the results have been included in the printed version of the paper.

Mr J W Figg observed that push-on fittings made of plastics are now being used for domestic systems made from stainless steel pipework. Dr Dulieu confessed that, as a metallurgist, he was prejudiced against plastics, but the important consideration would be their performance. Dr R A E Hooper of Arthur Lee and Sons felt that they might be satisfactory in low temperature applications.

Users Experience

Avoidance of Corrosion Problems

A. K. TILLER*

National Corrosion Service, Teddington, Middlesex, UK

ABSTRACT

Many factors must be considered in order to ensure the avoidance of corrosion problems in plumbed systems using potable water. Practical experience has shown that it is not sufficient to rely on good material selection. Attention must also be directed to the design of the installation, the quality of the water being handled, the operating conditions and the fabrication procedures, including the pre- and post-commissioning activities. All these can exert a dramatic influence on the subsequent performance of the system. Inadequate attention to any of these factors will eventually lead to costly corrosion problems, dissatisfied clients and customers, and the serious consideration of alternative materials which may still require to be assessed if satisfactory service is to be obtained.

1. INTRODUCTION

Those who suffer most from corrosion problems affecting potable water systems are frequently the industrial end users. Yet all too often, when industrialists seek guidance from the water companies on design and water quality aspects for the avoidance of corrosion problems, the response is limited to information about the chemical composition of the water being supplied. It is undoubtedly true that a prime requirement in combating corrosion is to 'know thine enemy', i.e. the

---

* The author is now an independent consultant, KT, Consultancy, Kingston-upon-Thames, Surrey, KT2 5BE, 081 546 7304.

chemistry of the water. However, this is not the only consideration and there is much that can be done at the design stage, as well as during the fabrication, installation, commissioning and operation of industrial installations to mitigate corrosion problems. This chapter summarises the key factors which need to be considered if serious corrosion problems are to be avoided in industrial installations using potable water.

2. EFFECTS OF WATER QUALITY

Figure 1 identifies some of the main features of water quality that can influence corrosion behaviour in potable water systems. Together with the pH value of the water, these control the likelihood of corrosion, depending on the materials included in the system. For instance, increasing hardness and alkalinity tend to suppress the corrosion of iron pipes whilst decreasing them will promote the dissolution of copper. The dissolved oxygen content of the water, through its ability to control the electrochemical potential of the metal and enable the development of differential aeration cells that can cause localised corrosion, is of particular importance. However, any of these factors, which include the presence of organic material, bacteria, and suspended solids, can dominate corrosion processes, depending on the particular circumstances. For example, biofilms formed by active bacteria adhering to the surface of the metal (in particular copper), can chelate copper ions, thus promoting the corrosion of that material. It is now recognised that bacteria and the presence of biofilms will encourage the corrosion of most of the metallic materials used in plumbed water systems. Hence, it is important that the end user should seek advice on how best to meet this particular challenge and to implement sound corrosion engineering practice which applies in each case. Such assistance is available from various government agencies, including the National Corrosion Service in Teddington, most of the water supply companies and numerous corrosion consultants.

3. DESIGN CONSIDERATIONS

It is the responsibility of the design engineer to ensure that appropriate measures are taken at the design stage to minimise corrosion problems overall during eventual use of the system. Figure 2 highlights the various aspects which should be addressed by the designer.

For example, the designer must be aware of the overall operating conditions and variations which are likely to arise. Knowledge of the effects of water

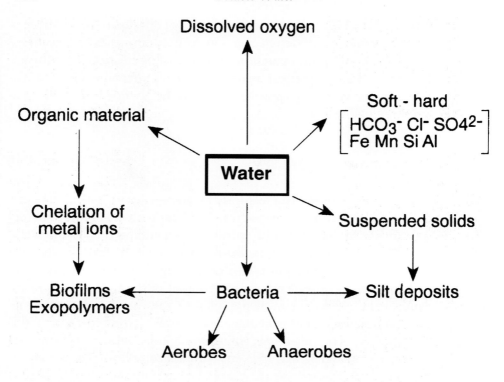

Fig. 1. Aspects of water quality that can influence corrosion in potable waters.

chemistry must be complemented with data on flow velocity, effect of temperature, fabrication aspects such as joints, bends, crevice regions to avoid erosion–corrosion, including the compatibility of the materials of construction.

The temperature of the system is another important consideration since different corrosion behaviour may be observed in hot water (at temperatures above 60°C) from that in water at lower temperatures.

During the materials selection process, compatibility between different metals is another important factor to be borne in mind. The need to avoid galvanic effects due to contact with electrochemically dissimilar metals should be obvious. However, galvanised pipes are still being used with copper components, with adverse consequences in some cases.

Because of the possibility with some materials of stress corrosion or corrosion fatigue cracking as a consequence of the conjoint effects of corrosion and static or alternating stresses, the designer must also take account of stress levels and types of stressing, including residual stresses, system stresses and operating stresses, in

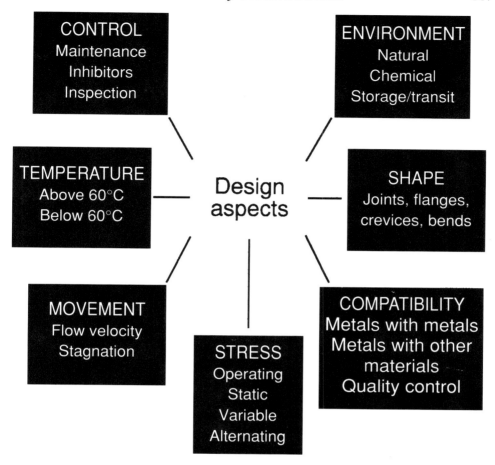

Fig. 2. Factors to which consideration must be given at the design stage.

association with data on cracking of the material under the environmental conditions to be encountered during service.

Finally, the designer should specify appropriate methods to control corrosion. These may include the periodic inspection and maintenance of the installation, and the incorporation of monitoring devices for corrosion, and/or water treatment procedures. The latter need to be monitored with equal vigilance.

4. MEASURES TO BE TAKEN AT SITE

A number of the pre- and post-constructional practices implemented on-site during the fabrication of industrial installations can affect subsequent corrosion behaviour, as indicated in Fig. 3. These include improper storage of pipework, since this can result in the admission of soil and other detritus, with consequential bacterial contamination. Also, adequate checks must be conducted on-site during fabrication to ensure that contamination by residual flux containing amines or chlorides does not occur. This is particular important in the case of copper and stainless steels.

Any stagnant water or dampness remaining after hydraulic testing can also cause corrosion problems. As in the case of the 'Scottish' problems with the pitting of copper tubing, it is recommended that systems should either be completely drained or maintained under flowing conditions after the hydrotest.

Disinfection practices such as BS 6700 are not easy to implement and can be abused, resulting, on occasion, in high levels of chlorine remaining in the system over the weekend due to inadequate checking. This can be particularly harmful for both copper and stainless steel. The temperature of the system during commissioning is also important since this can lead to a number of problems, not

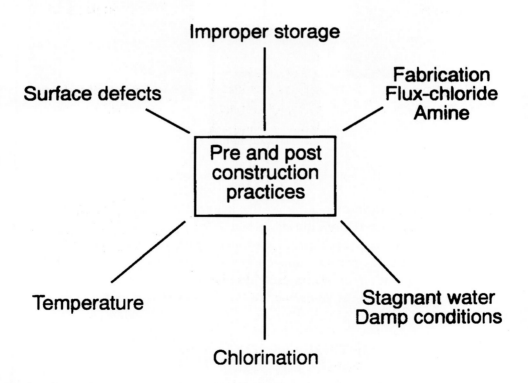

Fig. 3. On-site factors that can cause corrosion problems.

least simulating bacterial growth or selective corrosion such as stress cracking or even de-alloying phenomena.

The introduction of surface defects during fabrication must also be avoided. These can include score marks, and weld seams that are higher than normal. It is particularly important to confirm the acceptability of the internal surfaces of pipes.

5. OPERATIONAL FACTORS

Many aspects of the operating conditions to which plant is subjected can effect the likelihood of corrosion, as shown in Fig. 4. These include fluctuating oxygen levels which can vary between night and day, giving rise to differential aeration effects and elevated electrochemical potentials which promote corrosion. Intermittent use can have similar effects.

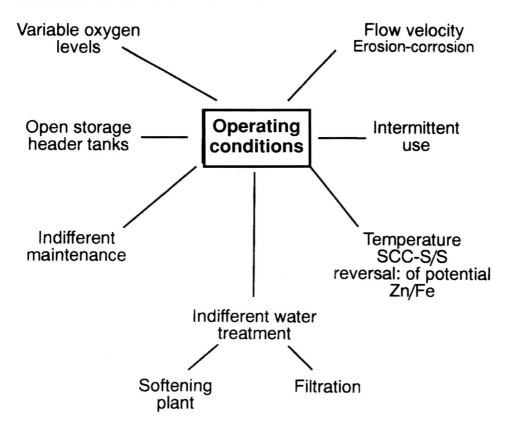

Fig. 4. Operating conditions that can impact on corrosion.

Open storage tanks can result in the admission of detritus to the system. Inadequate control of temperature can result in stress corrosion cracking of stainless steels or potential reversal as in systems fabricated in steel and galvanised steel.

Once the plant is operational it is essential that adequate maintenance of the system is achieved. This is particularly necessary when water treatment is adopted and that personnel understand the principles of such treatment procedures. Indifferent operation of softening plant allows slippage of chemical entities and filtration systems can lead to corrosion problems, as can indifferent maintenance where the need to remove pipes for inspection and to conduct water analyses is overlooked.

6. CONCLUSIONS

The avoidance of corrosion problems in industrial installations using potable water requires vigilance at all stages in the design, construction, operation and maintenance of plant. The user can only exert direct influence over the latter two aspects. It is the responsibility of designers and fabricators to do what they can to avoid subsequent problems with corrosion. The emphasis must be on adequate quality control at all stages.

Where assistance is required, it can be sought from Government bodies such as the National Corrosion Service, Water authorities, water treatment organisations, and/or suitably qualified consultants with appropriate experience.

DISCUSSION

Mr G G Page supported Mr Tiller's thesis that the importance of water purity needs to be emphasised. Recent experience in New Zealand has revealed problems with water systems containing very soft water that have not yet been generally recognised in the Northern hemisphere. He suggested that doubly distilled water should more properly be known as 'hellfire water'. Mr Tiller agreed that a tremendous grey area exists between chemists in water companies and those having to use the water. The users are provided with an analysis of the water and have to interpret it themselves. The fact that people do not want to use filters to remove suspended solids because they become contaminated means that particular attention has to be paid to the correct selection and use of materials. Deposit attack is still being experienced.

Mr H S Campbell commented that Mr Tiller's advice applied to large installations and asked what should be done to mitigate corrosion in domestic installations. Mr Tiller felt that the Institute of Plumbers should have responsibility to get the message across by, for example, informing the public of the type of problems; nitrite-based inhibitors can cause corrosion if a sufficiently high concentration is not maintained in central heating systems. Improved documentation for the public is the way forward.

Effects of Water Composition and Operating Conditions on the Corrosion Behaviour of Copper in Potable Water

H. SIEDLAREK, D. WAGNER, M. KROPP,
B. FÜSSINGER, I. HÄNSSEL AND W.R. FISCHER

*Märkische Fachhochschule, Laboratory of Corrosion Protection,
Frauenstuhlweg 31, 58590 Iserlohn, Germany*

1. INTRODUCTION

Decades of experience have shown copper to be a good material for potable water installations because of its good workability and corrosion behaviour. In Germany more than 60% of all tap water installations use copper.[1] Nevertheless, damage caused by pitting corrosion cannot be completely excluded in cold water where several adverse factors are acting at the same time.[2] Because of the complex dependence between the likelihood of pitting and the composition of the water, the condition of the metal surface, design considerations and operating conditions, the mechanism of this type of pitting is not clear at all. The dominant parameters within the water composition should be the anions[2a] and the pH.

This chapter will deal with the following questions:

(i) Do different anions within potable water influence the formation of layers of copper corrosion products in terms of their physical and electrochemical properties?

(ii) Does the pH of the potable water influence the formation of copper corrosion products on the electrode surface or within the electrolyte?

(iii) Is there a detectable effect of operating conditions, i.e. flow rate of the water and time of stagnation, on corrosion behaviour in potable water installations?

Potentiostatic measurements provide the best way of answering question (i).[3] Chloride is known to induce pitting in the copper/water system.[4,5] On the other hand it is mentioned in the literature that an exchange of sulphate ions for chloride ions has been used successfully as a countermeasure against the propagation of pitting of copper in cold water.[6] For these reasons, the first experiments were performed in chloride- and sulphate-containing electrolytes, the latter one forming neither complexes nor solid corrosion products with copper (I) ions. Galvanostatic experiments have been performed to answer questions (ii) and (iii) to guarantee a constant charge per unit time for the formation of copper corrosion products. The influence of operating conditions is being investigated in a laboratory loop under galvanostatic conditions.

2. EXPERIMENTAL

Potentiostatic experiments and galvanostatic experiments have been performed in a Faraday cage in the dark. Up to eight experiments can be performed in parallel applying different potentials or currents to the electrodes.

Measuring electrodes used for the experiments were prepared from hard copper tubes (SF-Cu, F37) according to DIN 1786[7] with a diameter of 22mm. Rings with an outer surface area of 8cm^2 were pre-treated with emery paper, polished with diamond paste and electropolished in ortho-phosphoric acid (70% w/w) with an anodic current density of $i = 0.2$A cm^{-2}. The rings were positioned in an electrode holder as shown in Fig. 1. Only the outer surfaces of these rings were polarised potentiostatically or galvanostatically using a common three electrode arrangement (Fig. 2). A ring of a platinum wire was positioned concentrically around the copper and used as a counter electrode. A mercury sulphate electrode (Hg/Hg$_2$SO$_4$/K$_2$SO$_4$ sat.) or a calomel electrode (Hg/Hg$_2$Cl$_2$/KCl sat.) was used as reference electrode. Electrode potentials were recalculated vs the standard hydrogen electrode (SHE).

A Haber–Luggin capillary was used to diminish the ohmic voltage drop. Movement of the electrolyte was established using a magnetic stirrer; if required, temperature of the electrolyte was kept constant at 20°C. The solutions were prepared with bidistilled water and chemicals of p. a. grade, i.e. for the sodium sulphate and sodium chloride electrolytes.

The electrolyte within the cell was exchanged at a rate of 1.5Ld^{-1} while performing the potentiostatic series. The electrolyte was aerated with a gas mixture of 20% vol. oxygen and 80% vol. nitrogen.

Since very dilute solutions were used and ohmic voltage drops could not be avoided completely, cut-off experiments were performed at the end of immersion. Electrode potentials shown in the figures correspond to the off potentials.

The electrodes were immersed for 140h during the potentiostatic series and for 24h in the galvanostatic experiments. The electrodes were subsequently pickled in 10 %w/w citric acid to determine the manifestations of corrosion according to DIN 50 905.[8] Pit depths were measured using a stereo microscope by differential

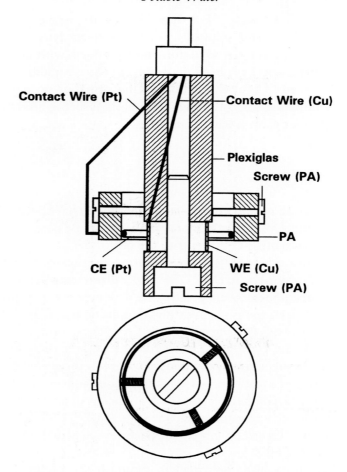

Fig. 1. Schematic diagram of an electrode holder.

focusing on the non-attacked surface and the bottom of what was obviously the deepest pit.

Galvanostatic experiments were conducted in laboratory loops as shown in Fig. 3 simulating different operating conditions. Flow velocities of 0, 0.08 and 0.6ms^{-1} could be established within these loops to simulate practical conditions within potable water installations. During these experiments the inner surface of the copper tube was polarised. The counter electrode, a platinum wire (0.5mm dia.) was passed coaxially through the copper tube and fixed with two polyamide holders. A stable Teflon tube (PTFE, i.d. = 1.5mm) formed the Haber-Luggin capillary and was located close to the inner surface of the copper electrode. This tube and the platinum wire were passed through the polyamide and sealed with silicone rings.

Effects of Water Composition on the Corrosion Behaviour of Copper 125

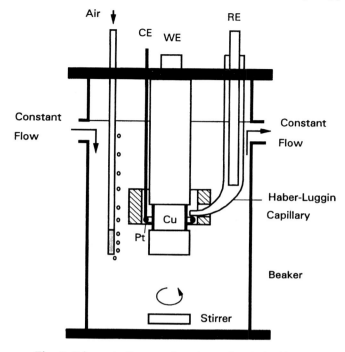

Fig. 2. Schematic diagram of an electrochemical cell.

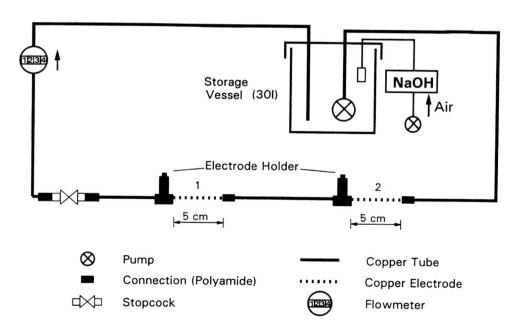

Fig. 3. Schematic diagram of a laboratory loop.

The experimental details are summarised in Table 1.

Table 1 Details of potentiostatic and galvanostatic experiments

Working Electrode (WE)	hard copper tube, 22 x 1 mm DIN 1786[7] (SF-Cu, F 37)
Electrode Area	outer surface: 8cm^2 inner surface: 32cm^2
Pre-treatment	potentiostatic series: polished, electropolished, prepolarised (10min at −800mV$_H$) galvanostatic series: polished, electropolished (outer surface) no pretreatment (inner surface)
Counter Electrode (CE)	platinum wire (0.5mm dia.)
Reference Electrode (RE)	Hg/Hg$_2$SO$_4$/K$_2$SO$_4$ sat. = 650mV$_H$ Hg/Hg$_2$Cl$_2$/KCl sat. = 242mV$_H$
Flow Velocity	potentiostatic series: < 0.1ms^{-1} galvanostatic series: stagnant (outer surface) stagnant, 0.08, 0.6ms^{-1} (inner surface)
Aeration	(i) laboratory air (80% N$_2$, 20% O$_2$) CO$_2$ removed: 2 NaOH + CO$_2$ → Na$_2$CO$_3$ + H$_2$O (ii) no aeration (iii) N$_2$-bubbling
Duration	potentiostatic series: 140h galvanostatic series: 24h
Electrolyte	NaCl, Na$_2$SO$_4$
Applied Potential (Potentiostatic series)	150–650mV$_H$
Applied Current (Galvanostatic series)	i = 1.25mA cm^{-2}

3. RESULTS

3.1 POTENTIOSTATIC SERIES IN SODIUM CHLORIDE ELECTROLYTES

The shape of the transients as shown in Fig. 4 is typical for measurements in sodium chloride solutions. A strong decrease of the current density can be observed even after a short polarisation time, indicating the formation of a protective layer consisting of corrosion products. This shape is independent of pH and the concentration of the salt. When the electrode is polarised cathodically, current densities remain very small; in the vicinity of the zero-current potential, current density may change its sign.

During anodic polarisation of copper in pure sodium chloride solutions a thin film of red to violet solid corrosion products is precipitated. Some brighter red spots are distributed randomly over the surface. Microscopic observation of these spots clearly shows that at these points the layer of corrosion products forms blisters. Above the blisters pores can be detected, and pits are observed within the layer and underneath the blister. Metallographic cross-sections show that the film covers the pits as stated by Lucey to be typical for pitting in cold water.[9] This film cracks and delaminates partially during drying (Fig. 5). It is obviously amorphous when it is wet.

Scanning Electron Microscopy (SEM) has shown that this film consists of extremely small crystallites after drying. However, in the area of the blisters the

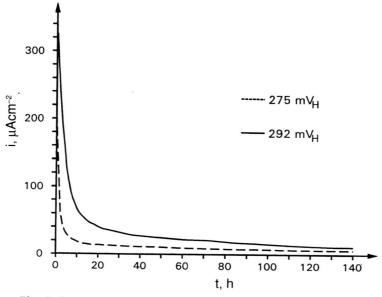

Fig. 4. Current–density transients measured in sodium chloride solution (C_{Cl-aq} = 200mg L^{-1}, pH = 8.5); parameter: off-potential.

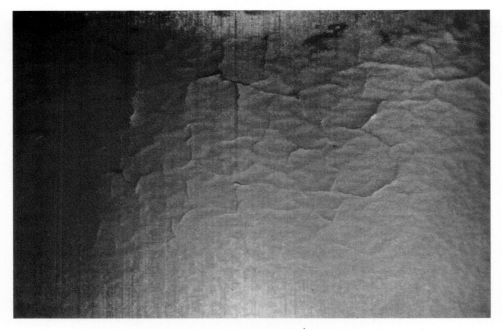

Fig. 5. Cu_2O layer obtained after drying after 140h polarisation at $200mV_H$ in 0.05 M NaCl electrolyte, magnification ×8.

crystals are larger. X-ray diffraction measurements show that this film consists of the two minerals: cuprite (Cu_2O) and nantokite (CuCl). Metallographic cross-sections and the dissolution of the oxide in diluted citric acid reveals that the total film consists of two layers: copper (I) chloride underneath copper (I) oxide as shown in Fig. 6. Furthermore, the pits described before can be seen on this cross-section.

To check whether the current transient is connected with a repassivation of the pits, the time of immersion was extended for one polarisation potential. In Fig. 7 the maximum pit depth is plotted vs time. The pit depths remain constant after *ca*. 100h polarisation time with pit depths of maximum $100\mu m \pm 30\mu m$, indicating the repassivation after this period of time.

3.2 POTENTIOSTATIC SERIES IN SODIUM SULPHATE ELECTROLYTES

Figure 8 shows current–density time curves for the anodic polarisation in sodium sulphate solutions for two potentials. Only a slight decrease of the current with increasing polarisation time can be observed. At these potentials a red–brown granular layer consisting of Cu_2O, as shown by X-Ray Diffraction Analysis (XRDA), can be observed. This is depicted in Fig. 9 (p.131) for a copper surface exposed at $550mV_H$ in 0.05 M Na_2SO_4 after 140h polarisation time (magnification

Effects of Water Composition on the Corrosion Behaviour of Copper

Fig. 6. Cross-section of an electrode polarised at 300mV$_H$ for 140h in 180mg L^{-1} NaCl solutions (magnification × 64).

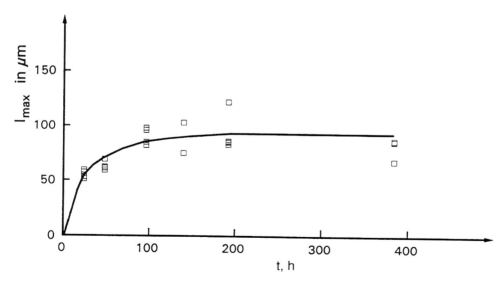

Fig. 7. Maximum pit depth as a function of polarisation time measured in sodium chloride solution (C$_{Cl-aq}$ = 200mg L^{-1}, U = 400mV$_H$).

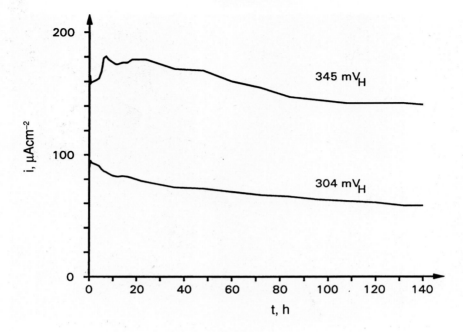

Fig. 8. Current–density transients measured in sodium sulphate solution ($SO_4^{2-}{}_{aq}$ = 200mg L^{-1}, pH = 8.5); parameter: off-potential.

×64). A cross-section of an electrode, after being polarised at 450mV$_H$ for 140h in $0.5 \cdot 10^{-3}$ M Na$_2$SO$_4$ is shown in Fig. 10. The only manifestation of corrosion that can be detected underneath the granular Cu$_2$O layer is general attack. Investigation of the granular Cu$_2$O layer with SEM shows that it consists of an assembly of big cuprite crystals containing a large amount of free volume. This badly adherent layer develops cracks during drying and falls off the surface. No influence of electrolyte concentration and pH in the range of 4 < pH < 10 on the formation of reaction layers on the electrode surface and the manifestation of corrosion can be detected.

3.3 GALVANOSTATIC SERIES IN SODIUM SULPHATE ELECTROLYTES

3.3.1 Corrosion products within the electrolyte

One further result from the potentiostatic series in sodium sulphate electrolytes has to be mentioned: only cuprite was found on the electrode surface, whilst in a copper tube installation operating with a sulphate rich potable water, cuprite and basic copper sulphates, e.g. posniakite, can be detected on the tube surface. In special cases black copper (II)-oxide is also found.[10] This led to the investigation of the copper corrosion products within the electrolyte in more detail, with special regard to pH and stirring effects.

Fig. 9. Granular red–brown Cu_2O layer on a copper surface obtained after 140h polarisation at 550mV in 0.05 M Na_2SO_4 solution (magnification × 64).

Fig. 10. Cross-section of a copper electrode exposed in 0.5×10^{-3} M Na_2SO_4 at pH 4 at $450mV_H$ for 140h, magnification × 64 (tube thickness 1mm).

These experiments were performed in the set-up shown in Figs 1 and 2 with different electrolyte concentrations in the absence and presence of light. The corrosion products were rated qualitatively according to their colour and were identified by XRDA. The detailed experimental conditions and results are shown in Table 2.

Table 2 Experimental conditions and results of galvanostatic experiments

Electrolyte concentration	Aeration	pH t = 0h	pH t = 24h	Corrosion products colour	Stirring
DARK					
1 M	80%N_2/20%O_2	8.7	10.5	brown–black	-
1 M	80%N_2/20%O_2	8.7	9.8	brown–black	+
1 × 10³ M	N_2	7.0	10.5	brown–black	-
1 × 10³ M	N_2	7.0	8.0	brown–black	+
1 M	no aeration	4.0	n.d.	blue	-
1 M	no aeration	4.0	n.d.	blue	+
1 × 10³ M	no aeration	6.5	n.d.	brown–black	+
1 × 10³ M	80%N_2/20%O_2	6.5	n.d.	brown–black	-
1 × 10³ M	no aeration	6.5	n.d.	brown–black	-
LIGHT					
1 × 10³ M	80%N_2/20%O_2	6.5	n.d.	brown–black	+
1 × 10³ M	N_2	6.5	n.d.	brown–black	+
1 × 10³ M	N_2	6.5	n.d.	brown–black	-
1 × 10³ M	no aeration	7.0	n.d.	brown–black	+

n.d. = not determined

The results that can be derived from these measurements, can be summarised as follows:

(i) No influence of stirring on the formation of corrosion products within the electrolyte can be detected.

Effects of Water Composition on the Corrosion Behaviour of Copper

(ii) The same holds for the different kinds of aeration.
(iii) The measurements in the absence and presence of light yield the same results.
(iv) The pH values after performing the experiments in the absence of light are higher than at the start. The pH values after performing the experiments in the light have not been measured.
(v) All measurements that started in the neutral pH range (6.5–8.7) yielded brown–black corrosion products. Blue corrosion products were only obtained in the experiments that started at pH 4.

For a more detailed view, the same galvanostatic experiments have been performed in 10^{-3} M Na_2SO_4 without aeration, separating the anodic and cathodic partial reaction by the use of two electrochemical cells connected via an agar–agar salt bridge. These experiments were started with two different pH-values (4 and 11) that were adjusted within the cell with the anodic partial reaction. The results are summarised in Table 3.

Table 3 Results of galvanostatic experiments in the cell with the anodic partial reaction

	pH (t = 0h)	pH (t = 24h)	Corrosion products	Colour electrolyte
1×10^3 Na_2SO_4	4	5	none	blue
1×10^3 Na_2SO_4	11	4	none	blue

The following results were obtained:

(i) The pH remained nearly constant in the acidified electrolyte but in the alkalised cell a strong acidification to pH 4 could be observed.
(ii) No corrosion products were obtained in either case, only blue-coloured electrolytes.

The copper ions in these cells were precipitated in two different ways:

(i) The solution is alkalised directly to pH 11 under stirring. A blue deposit is obtained that changes its colour to brown–black within a few minutes.
(ii) The solution is alkalised slowly to pH 5.6 under stirring. A blue precipitate is obtained that has not changed its colour any more.

This blue corrosion product and three characteristic brown–black samples in Table 2 were investigated by XRDA.

The brown–black corrosion products consist of Cu°, Cu_2O and CuO. The brown–black colour is caused by the CuO.[11] The main component of the blue corrosion product can be identified as posniakite ($CuSO_4 \cdot 3Cu(OH)_2$).

3.3.2 Adhesion of corrosion products on the electrode surface
During the performance of the galvanostatic series described in the preceding section, a red–brown granular layer consisting of Cu_2O with non-protective properties was obtained, independent of the experimental conditions described in Table 2. One significant difference was found concerning the influence of stirring. Under stirring conditions a non-adherent granular Cu_2O layer was obtained whilst copper electrodes polarised under non-stirring conditions showed an adherent Cu_2O layer. To check this influence in more detail, galvanostatic experiments were performed in a small laboratory loop as shown in Fig. 3, using different flow velocities, v. The aeration of the electrolyte within this loop was performed within the 30L storage vessel. The adhesion of the corrosion products to the electrode surface was tested with a transparent adhesive tape. The results are shown in Fig. 11 and confirm the results obtained in the electrochemical cells. Under stagnant conditions (v = $0 ms^{-1}$) an adherent Cu_2O layer forms on the electrode surface whilst measurements under constant flow conditions (v = $0.6 ms^{-1}$) yield a non-adherent Cu_2O layer.

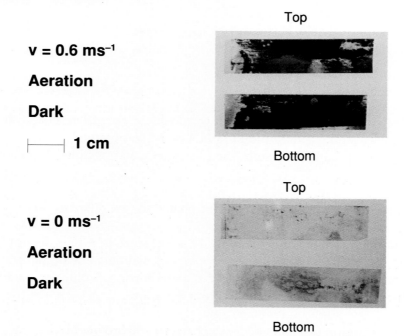

Fig. 11. Adhesion tests of the Cu_2O layer with a transparent tape as a function of flow conditions.

4. DISCUSSION

4.1 POTENTIOSTATIC SERIES IN SODIUM CHLORIDE AND SODIUM SULPHATE ELECTROLYTES

It can be deduced that in a sodium chloride electrolyte an adherent protective film of Cu_2O is formed via the hydrolysis of Cu (I) Cl:

$$Cu \rightarrow Cu_{aq}^+ + e^- \quad (1)$$

$$2\, Cu_{aq}^+ + 2\, Cl_{aq}^- \rightleftharpoons 2\, CuCl \quad (2)$$

$$2\, CuCl + H_2O \rightleftharpoons Cu_2O + 2H^+ + 2Cl^- \quad (3)$$

Within the pits an adherent layer of copper (I) chloride and big crystals of copper (I) oxide are formed. This type of pitting is a repassivating pitting. The maximum pit depth depends on the corrosion potential, but the observed values do not exceed 150 μm even after a prolonged polarisation time.

In sodium sulphate electrolytes the granular, non-protective Cu_2O layer is formed via the following reactions:

$$Cu \rightarrow Cu_{aq}^+ + e^- \quad (4)$$

$$2\, Cu_{aq}^+ + H_2O \rightleftharpoons Cu_2O + 2H^+ \quad (5)$$

These results point out that the corrosion behaviour of copper in cold water is drastically influenced by the chloride ions. During anodic polarisation in sodium chloride solutions the protective film of copper (I) oxide is formed via hydrolysis of the primary product, copper (I) chloride, whilst the assembly of copper (I) oxide crystals is deposited via the direct reaction of copper (I) ions with water.

In sodium sulphate solutions a non-protective crystalline layer of copper (I) oxide is deposited during anodic polarisation of the electrode. This layer is built via direct reaction of cuprous ions with water. This granular layer does not inhibit the anodic oxidation of the copper.

The results show clearly that the corrosion behaviour of copper in cold water is drastically influenced by its anionic composition:

(i) a sulphate-rich water causes general attack and does not inhibit the anodic partial reaction, and
(ii) a chloride-rich water induces pitting but inhibits pit penetration.

4.2 Galvanostatic series in sodium sulphate electrolytes

At lower pH-values mainly blue posniakite ($CuSO_4 \cdot 3Cu(OH)_2$) is formed as a copper (II)-corrosion product within the electrolyte whilst higher pH-values yield the formation of brown–black copper (II)-oxide. Within the latter electrolyte copper(O) and copper (I)-oxide could be identified, too, due to the reduction of copper ions formed at the counter electrode to copper (O) due to the fact that the granular Cu_2O formed on the electrode surface falls off into the electrolyte during prolonged galvanostatic polarisation.

The observation that two different corrosion products are formed at different pH values can be explained by thermodynamic considerations. Figure 12 shows the solubility of copper ions as a function of pH with respect to the formation of CuO or $CuSO_4 \cdot 3Cu(OH)_2$ within the electrolyte for different sulphate ion concentrations based on solubility products given in the literature: (pK (CuO) = −22;[12] pK (CuO) = −20.12;[13] pK ($CuSO_4 \cdot 3Cu(OH)_2$ = −17.13.[14] Figure 12 is based on the values given by Feitknecht[13] and Näsänen.[14] The formation of $Cu(OH)_2$ formed first at higher pH-values is not considered in this thermodynamic treatment because it is transformed into the stable black–brown Cu(II)O within a short period of time.[15] The horizontal line in Fig. 12 shows the calculated concentration of Cu^{2+} ions that would be obtained within the electrolyte (vol. 1.2L) after 24h galvanostatic polarisation with a current density of 1.25mA cm^{-2} with the assumption that only Cu^{2+} ions would be formed. It can be derived from this figure that soluble copper (II)-ions should be formed at low pH, posniakite at medium pH and copper (II)-oxide at high pH.

In 1 M Na_2SO_4 electrolytes posniakite is stable in the range of 4.2 < pH < 8.2, and in 1 × 10^{-3} M solutions in the range of 4.7 < pH < 6.8. Based on these considerations, the results shown in Tables 2 and 3 concerning the formation of corrosion products within the electrolyte can be explained.

A slow alkalisation of a 1 × 10^{-3} M Na_2SO_4 solution (pCu^{2+} = 2.43) under strong stirring yielded a stable blue deposit consisting of posniakite according to the literature.[16] A fast deposition would lead to a green colloidal alkaline copper sulphate that changes its physical properties with time.[16] The fast transition of blue $Cu(OH)_2$ to brown–black CuO described in Ref. 15 is also verified.

A further important observation must be mentioned. Under stagnant conditions an adherent granular Cu_2O layer is formed whilst under constant flow conditions the formation of a non-adherent layer can be observed. It seems that the pH value at the electrode/electrolyte phase boundary plays a role concerning the morphology of this layer. The H$^+$ ions are produced during its formation. This would mean that under stagnant conditions the pH is lower at the phase boundary because of a large diffusion layer and an adherent crystalline Cu_2O layer is formed

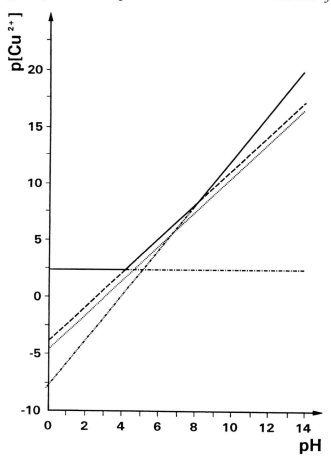

Fig. 12. Solubility of copper ions as a function of pH with respect to the formation of CuO or $CuSO_4 \cdot 3Cu(OH)_2$ within the electrolyte for different sulphate concentrations.

———	smallest solubility of Cu^{2+} ions
.. — .. —	CuO
....	$CuSO_4 \cdot 3Cu(OH)_2$, 1×10^{-3} M Na_2SO_4
— — — — —	$CuSO_4 \cdot 3Cu(OH)_2$, 1 M Na_2SO_4
— · — · — ·	Cu^{2+}

whilst under stirring conditions the pH is higher leading to the formation of an amorphous layer of corrosion products that can be removed with an adhesive tape. More detailed investigations concerning this aspect are currently in progress.

5. CONCLUSIONS

The observation that Cu_2O layers adhere so differently in different types of potable water makes it probable that the flow velocities within a copper pipe distribution system, i.e. the operating conditions (stagnant, intermittent, constant flow) influence the corrosion propensity the system. The corrosion propensity is a function of the kind of Cu_2O layer, it is high with a sulphate-type layer and low with a chloride-type layer.

The results that different copper (II)-corrosion products are formed within the electrolyte can be confirmed by observations in practice where copper (II)-oxide and posniakite could be identified in damaged copper tubes of an installation system operating with potable water that contains mainly sulphate as an anion.[10] Posniakite occurred in the area where the anodic partial reaction leading to pitting took place and copper (II)-oxide in the area where the cathodic partial reaction showing only general attack took place.[10]

6. ACKNOWLEDGEMENT

We would like to thank the International Copper Association (ICA) and the German Ministry of Research and Technolgy for financial funding. Furthermore, parts of this work are supported by the European Community (BRITE/EURAM project no. 4088).

REFERENCES

1. A. Baukloh, H. Protzer, U. Reiter and B. Winkler, *Metall.*, 1989, 43, 26.
2. DIN 50 930, part 5; "Corrosion of metals; corrosion behaviour of metallic materials against water; scale for evaluation for copper and copper alloys", Beuth-Verlag, Berlin, 12/1980.
2a. W. Fischer, B. Füßinger and I. Hänßel, "Influence of anions on the pitting behaviour of copper in potable water", Proceedings of 12th Scandinavian Corrosion Congress & Eurocorr '92, pp. 769-778.
3. W. R. Fischer and W. Schwenk, in *Electrochemical Corrosion Testing* (E. Heitz, J. C. Bowlands, F. Mansfeld, eds), DECHEMA Monographien 1985, 101, 87.
4. F. M. Al-Kharafi and Y.A. El-Tantawy, *Corros. Sci.*, 1982, 22, 1.
5. R. May, *J. Inst. Metals* 1954, 32, 65.
6. C.-L. Kruse and P.-K.-J. Enzenauer, "Korrosionsschutz in Trinkwasserleitungen der Hausinstallation durch Veränderung der Wasserbeschaffenheit mit Anionenaustauschern"; Sanitar- und Heizungstechnik Vol. 2, 1987.
7. DIN 1786, "Copper tubes for plumbing, seamless drawn", Beuth-Verlag Berlin, 05/1980.

8. DIN 50 905, "Corrosion of metals, corrosion testing, general guidance", 09/1985.
9. V. F. Lucey, *Werkstoffe und Korrosion* 1975, 26, 185.
10. W. R. Fischer, H. H. Paradies, D. Wagner and I. Hänßel, "Copper deterioration in a water distribution system of a county hospital in Germany caused by microbially induced corrosion - I. Description of the problem", *Werkstoffe und Korrosion* 1992, 43, 56–62.
11. D'Ans-Lax, *Taschenbuch für Chemiker und Physiker, Band 1, 3. Aufl.*, Springer-Verlag, Berlin, Heidelberg, New York, 1967.
12. M. Pourbaix, *Atlas of Electrochemical Equilibria in Aqueous Solutions*, Pergamon Press Ltd., 1967, pp. 384.
13. W. Feitknecht; "Über die Löslichkeitsprodukte der Oxide und des Hydroxids von Kupfer und über die Löslichkeit von Kupferhydroxid in Natronlauge", *Helv. Chim. Acta*, 1944, 27, 771-5.
14. R. Näsänen, "The effect of complex formation between cupric and sulphate ions on the equilibrium on cupric trihydroxysulphate in mixed aqueous solutions of cupric and potassium sulphate", *Acta Chem. Scand.* 1949, 3, 1400–1404.
15. *Gmelins Handbuch der anorganischen Chemie*, Kupfer, S.N. 60, 8th ed; Verlag Chemie, 1958, Part B, p. 101.
16. H. T. S. Britton, *J. Chem. Soc.* 1925, 127, 2148/59, 2151, 2796/2807, 2798.

DISCUSSION

Mr G G Page commented that the corrosion mechanisms described by Dr Wagner for copper in cold water environments differed from the one he had himself reported in New Zealand. However, Dr Wagner thought it unlikely that any single mechanism could account for all observed corrosion processes.

Mr A K Tiller asked what flow rate would be required to produce horseshoe-shaped corrosion due to erosion–corrosion. Dr Wagner stated that this was *ca.* 0.6 m s^{-1}, i.e. higher than could be achieved with a magnetic stirrer or small pump. Mr A J Graham of ERA Technology followed up with a question about the relationship between the time taken for surface film breakdown to occur and the flow velocity. The author replied that stirring appeared to have little effect on the surface layer which was entirely composed of Cu_2O. Mr J Figg asked how amorphous deposits had been identified by X-ray diffraction. Dr Wagner replied that the deposits, which were amorphous when wet, had crystallised during the drying process. This led Mr Figg to conclude that the deposit must have been the hydroxide, rather than the oxide, when wet.

Observations on the Corrosion Behaviour of Copper in Glasgow Tap Water

T. HODGKIESS AND J. AKHTAR*

Department of Mechanical Engineering, University of Glasgow, Scotland

ABSTRACT

This paper describes some experimental work focused on the electrochemical behaviour of copper in Glasgow tap water with the objective of providing information that might contribute to the understanding of recent corrosion problems of copper pipes in Scottish public-water systems.

Most of the work has involved carrying out DC-polarisation tests on small specimens taken from two sources of copper pipes. Copper exposed in experiments conducted for periods of up to a year at ambient temperature, has exhibited passive characteristics. Similar behaviour has also been observed at temperatures of up to about 50°C. However, at temperatures of around 60°C, the electrochemical tests have not revealed systematic evidence of similar passivity but have provided indications that the resistance of the material to significant corrosive attack, in the circumstances investigated, may be inherently marginal at these higher temperatures. Other experiments have involved an assessment of the influence of high levels of natural and synthetic organic acids and of sodium hypochlorite on the corrosion process.

1. INTRODUCTION

In the present century, copper has become established as the major material for building services, water-distribution systems and enormous quantities of copper pipe have been installed worldwide.[1] In the last 30 years, problems due to

---

*Present address: Natural Gas Power Station, Wapala, Piramghaib Mulfaw, Pakistan.

corrosion of such copper pipe systems have arisen from time to time and have received considerable description and discussion in the literature including periodic reviews.[1-3]

In recent years, there have been major problems with pitting corrosion of copper water distribution pipes in Scotland. The problems have been most widely publicised in relation to hospitals because numerous such installations have been affected[4] with instances of complete renewal of all cold-water and hot-water piping being necessitated at enormous costs — figures of £6 million being quoted[4] for one hospital near Glasgow.

However, the problem is not isolated to hospitals. A few years ago, after the problem had started to appear in hospitals, a survey[5] was undertaken of a number of organisations in the West of Scotland. This questionnaire confirmed the widespread existence of copper pipe corrosion in local hospitals but also provided the following data from the large hotels which were approached:

- Number of hotels approached = 36
- Number of hotels returning questionnaire = 27
- Number of hotels reporting corrosion problems of copper pipework = 9.

Schools, a prison and individual dwellings have also been affected. Figure 1 shows pitting attack that caused leakages in a vertical rising main in a Glasgow tenement dwelling. The occurrence of rows of pits, evident in Fig. 1, is often a noticeable feature in corroded pipework removed from buildings in Scotland.

In the past few years, extensive investigations costing around £1 million,[4] have been initiated into the so-called 'Scottish problem' without apparently as yet being able to provide a full understanding of the corrosion processes. These activities have involved:

(i) examinations of corroded pipework from Scottish hospitals and comparison of the observed features with those associated with previous outbreaks of corrosion in other countries,
(ii) corrosion monitoring schemes in Scottish hospitals,
(iii) related research studies — mostly undertaken outside Scotland.

The work described herein represented a limited, unfounded but sharply-focused study of the electrochemical behaviour of copper pipe material when exposed to Glasgow tap water. The objective was to secure information on the corrosion behaviour of copper in relevant conditions as a contribution to the overall state of knowledge from which a fuller understanding of the pitting corrosion phenomenon might hopefully emerge.

Fig. 1. Pitting in copper pipe from a building in Glasgow.

2. EXPERIMENTAL

Most of the experiments described herein were conducted on specimens, of approximately 1 cm × 1 cm surface area, cut from two samples of copper pipe:

(1) pipe conforming to BS2871, Table X,
(2) pipe exhibiting severe pitting which had been removed from a large Glasgow hotel.

The aim of studying the behaviour of the second pipe material was to ascertain if there was anything uniquely poor about the corrosion resistance of this pipe material. The specimens taken from this second batch of pipe were carefully removed from areas remote from pits. Electrical connecting wires were attached to the specimens followed by the casting of epoxy resin around the specimen and abrasion of the exposed face to 600 grit, washing, degreasing and drying.

The experiments involved measurement of the free corrosion potential, E_{corr}, of the specimens and also carrying out DC polarisation tests on separate samples. The main series of these latter tests concentrated on the anodic polarisation characteristics but some cathodic polarisation scans were also made. The specimen configuration was designed to facilitate electrochemical measurements in the low conductivity environment presented by Glasgow water: this involved the arrangement depicted in Fig. 2 in which a small reproducible distance was maintained between the copper working electrode and the counter electrode (titanium) Electrode potentials were measured using a saturated calomel electrode (SCE). Specimens were exposed for varying periods of up to about one year with polarisation scans always being undertaken on previously unpolarised specimens.

The exposure medium for most of the work was water from the normal water supply system in the Glasgow University laboratories. The intention was to provide a gently replenished water supply for these experiments and this was achieved by placing batches of specimens in ten-litre buckets with the water from the tap continually but slowly flowing into the bottom of the bucket and gently overflowing to waste. The tests involved exposure of specimens to:

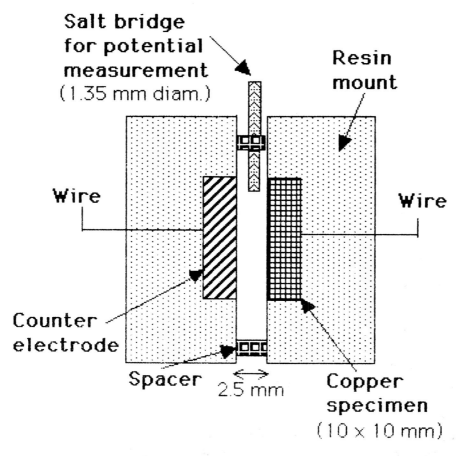

Fig. 2. The compact cell system.

(1) cold tap water (temperatures around 10°C),
(2) tap water direct from the hot-water tap (at temperatures between 40–50°C throughout the duration of the experiments),
(3) tap water direct from the hot-water tap but boosted to 60–65°C by the use of a titanium immersion heater located in the bucket.

The Glasgow public water, in common with much of the public water supply in Scotland, is very low in dissolved salts (TDS *ca.* 20ppm) and in alkalinity (< 10ppm) and is extremely soft (*ca.* 4ppm Ca). It can vary appreciably in pH, 6.6–9.0. Indeed, another facet of this work involved measurements of the trend with time of the pH of Glasgow tap water.

In addition to the main series of experiments using Glasgow tap water, a few tests were carried out in other environments (listed below) to assess the influence of the presence of organic matter (including organic acids) and also the effects of high levels of chlorination on the corrosion behaviour:

(1) natural water sampled from a loch near to Glasgow,
(2) distilled water containing 50ppm Gallic acid; this was utilised because the extremely complex humic and fulvic acids present in the natural water found in Scotland are considered to comprise repeating arrays of gallic acid molecular groupings,
(3) Water containing 50ppm sodium hypochlorite, NaOCl,
(4) Water containing 50ppm Gallic acid + 50ppm NaOCl.

The experiments in the above four solutions were carried out with copper specimens exposed to the water in beakers in static conditions.

3. RESULTS

3.1 ANODIC POLARISATION IN COLD, MOVING TAP WATER

Figure 3 shows the results of a series of anodic polarisation scans made on separate specimens during the initial 9 days of exposure of the copper from the hotel pipework. (Note that, in all the results graphs, this material is designated by the label, 'Copper' with the other source of material studied being labelled, 'BS 2871'.) As in all the polarisation tests, the specimens were shifted from their natural free-corrosion electrode potentials at a rate of 15mV min^{-1}, and the currents flowing in the electrochemical circuit were monitored. When a specimen is corroding actively (i.e. at significant rates) in the unpolarised state. rapidly increasing currents are recorded as the potential is shifted in the positive direction from the free corrosion value (E_{corr}). On the other hand, the presence of a passive surface film (conferring corrosion protection to the specimen) results in the passage of only tiny currents (a few μA or less) as the potential is positively scanned. Such passive behaviour is obviously indicated by the results in Fig. 3 which reveals that there was a substantial range of potentials over which passivity was maintained on the copper before the currents commenced to increase at the faster rates indicative of the breakdown of the protective surface film.

Although difficult to observe clearly from Fig. 3, the free corrosion potential was in fact moving in the positive direction with passage of time and this general positive drift in E_{corr} is shown for the two sources of copper in Fig. 4. There were some clear fluctuations in potentials including one substantial one after 14d for the 'copper' but the general trend was clearly in the positive direction.

The two sources of copper used in this programme both exhibited passive behaviour but with some interesting differences, especially in the early days of exposure — as illustrated in Figs 5 and 6. After 1 day, the free corrosion potential of the copper of known specification was much more negative but the other material (which had pitted in service) actually appeared to possess a more protective film — as evidenced by the smaller polarisation currents above E_{corr}. These differences between the two materials were still evident after 6d but, at later stages of exposure typified by the 105-day results in Fig. 5, the free corrosion potentials and the anodic polarisation characteristics of the two materials were much more similar.

A more detailed indication of the effect of exposure time on the anodic polarisation behaviour is provided in Figs 7 and 8. The main message is clearly one of the definite maintenance of corrosion protection for the period, in excess of one year, studied in this research work.

The appearance of the specimens after extended exposures, without being subjected to any polarisation tests, was of a thin uniform brownish-coloured surface film but containing a silt-like deposit.

3.2 Anodic polarisation in moving tap water at 40–50°C

The behaviour in this heated tap water was similar to that in cold water. Thus the free corrosion potentials were observed to move significantly in the positive direction in the early days of exposure but, thereafter, any continued drift was much less noticeable. Occasional fluctuations in E_{corr} were recorded (like those, shown in Fig. 4, observed in cold water) but with no evidence that such fluctuations were associated with changes in corrosion protection of the copper. Indeed, all the anodic polarisation scans provided evidence of passivity over the approx. one-year duration of the exposures. This point is demonstrated for the BS 2871 copper in Fig. 9 and similar behaviour was observed for the other source of copper.

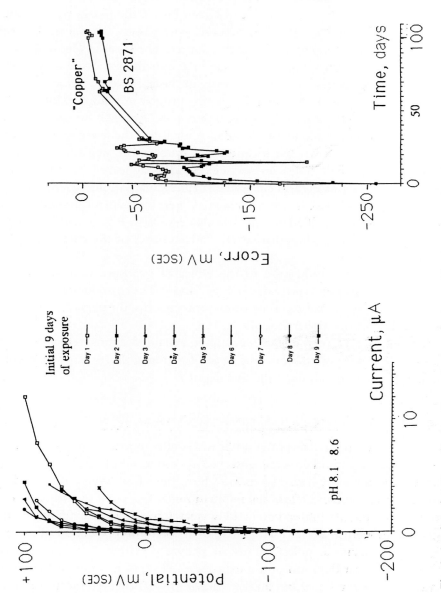

Fig. 3. Anodic polarisation of 'copper' in moving tap water.

Fig. 4. Free corrosion potential in moving cold tap water.

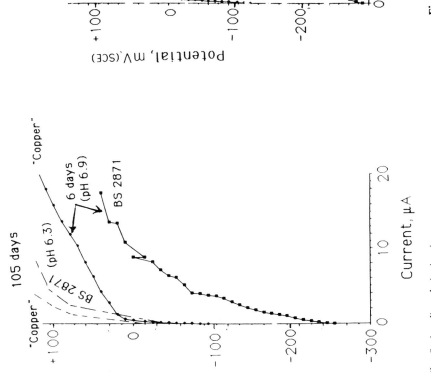

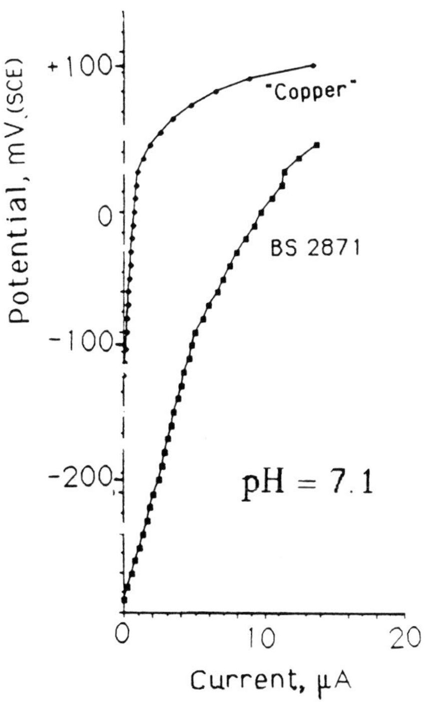

Fig. 6. Anodic polarisation after 1d in moving cold tap water.

Fig. 5. Anodic polarisation in moving cold tap water.

105 days
405 days
9 days
1 day
6 days
+100
0
-100
-200
Potential, mV (SCE)
E_{corr} values:-
1 day, -123 mV
6 days, -95 mV
9 days, -78 mV
105 days, -12 mV
405 days, -41 mV
0
10
20
Current, µA

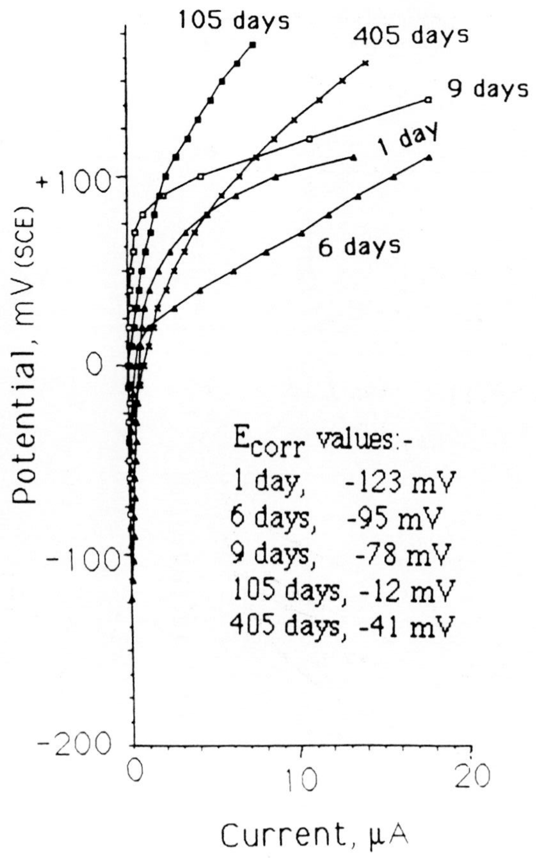

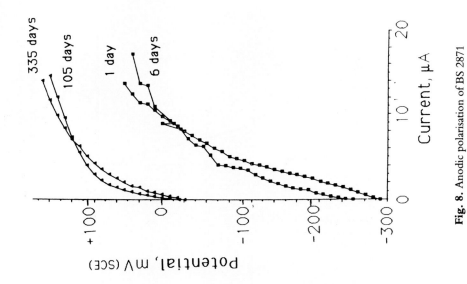

Fig. 7. Anodic polarisation of 'copper' in moving cold tap water.

335 days
105 days
1 day
6 days
+100
0
-100
-200
-300
Potential, mV (SCE)
0
10
20
Current, µA

Fig. 8. Anodic polarisation of BS 2871 copper in moving cold tap water.

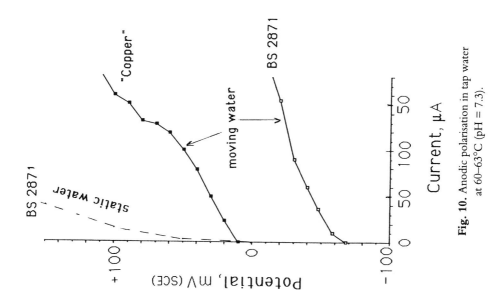

Fig. 10. Anodic polarisation in tap water at 60–63°C (pH = 7.3).

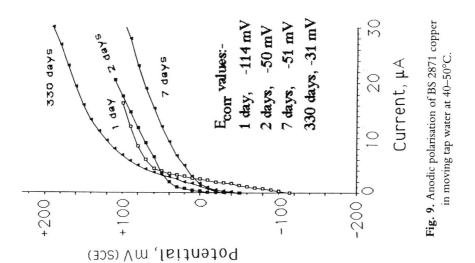

Fig. 9. Anodic polarisation of BS 2871 copper in moving tap water at 40–50°C.

The appearance of the specimens after long exposures to this hot water was similar to those exposed to cold tap water.

3.3 Anodic polarisation in moving tap water at 60–65°C.

Upon immediate exposure in this, higher temperature moving tap water, the anodic polarisation results were not indicative of passive behaviour. This is shown clearly by the rapidly increasing currents plotted on Fig. 10. (However, as also shown in Fig. 10, in *static* water, passive behaviour was apparent.)

The free corrosion potentials moved in the positive direction in the early days of exposure (Fig. 11), but after a few months' exposure, there were some quite large variations in E_{corr}.

The non-passive behaviour shown upon immediate exposure appeared to be maintained throughout the extended immersion periods (Figs 12, 13). The anodic polarisation curves indicating active corrosion were further analysed in the conventional manner by replotting in semi-logarithmic form (E vs log I) and extrapolation of the linear portion of the graph to yield corrosion current, I_{corr}. Further calculations using Faraday's laws, yielded low metal losses of less than 0.1 mm/year over the period 1 day to 1 year.

Specimens exposed for a year without undergoing any polarisation showed some indications of the initiation of tiny corrosion pits.

3.4 Anodic polarisation in 50ppm sodium hypochlorite solution

Upon immediate exposure to this highly-chlorinated solution at 18°C, anodic polarisation scans (Fig. 14) yielded high currents indicative of non-passive behaviour. However, after 32d in this medium, the usual cold water passive polarisation result was obtained. Moreover, at 60°C, a similar low current plot was recorded (Fig. 14) even after first immersion. The change from active to passive behaviour at 18°C is considered to be likely to be due to the gradual release of the chlorine from the water during the 32d exposure and the low current response at 60°C could well have been due to similar, rapid evaporation of chlorine during the short period before and during the polarisation test.

3.5 Anodic polarisation in natural, loch water

Typical of many waters in Scottish lochs, this water was highly coloured with the peaty colour due to the presence of the naturally present organic compounds. This water was also of low pH (3.5–4.0) and (Figs 15, 16) promoted non-passive indications on immediate exposure at 18 and 60°C. However, after 72d immersion, the pH had risen to above 6 and anodic polarisation scans produced low current responses.

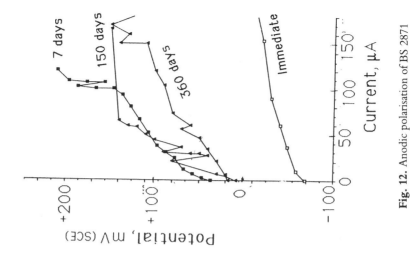

Fig. 12. Anodic polarisation of BS 2871 copper in tap water at 60–65°C.

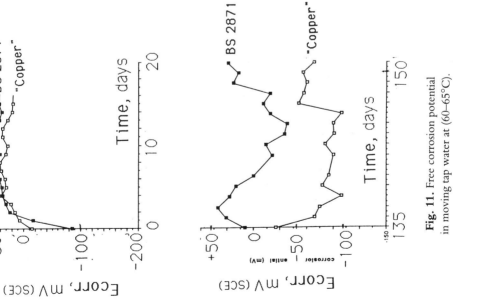

Fig. 11. Free corrosion potential in moving tap water at (60–65°C).

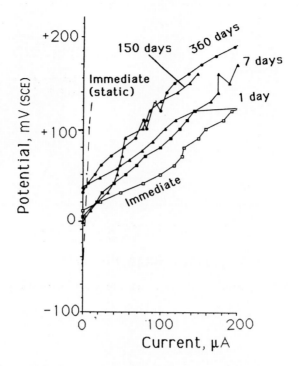

Fig. 13. Anodic polarisation 'copper' in tap water at 60–65°C.

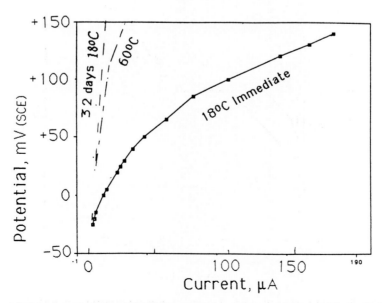

Fig. 14. Anodic polarisation BS 2871 copper in static water containing 50ppm NaClO.

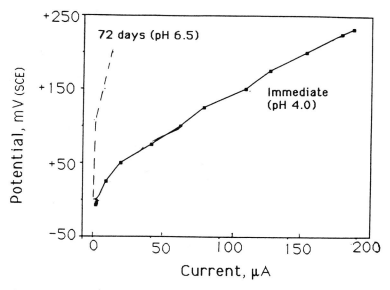

Fig. 15. Anodic polarisation BS 2871 copper in static, natural water at 18°C.

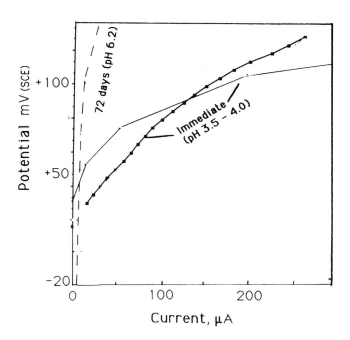

Fig. 15. Anodic polarisation BS 2871 copper in static, natural water at 60°C.

3.6 ANODIC POLARISATION IN WATER CONTAINING 50PPM GALLIC ACID

Figure 17 shows several anodic polarisation curves obtained in water containing gallic acid. Upon initial exposure in the low-pH gallic acid solution at 18°C, the polarisation currents were relatively high (i.e. not characteristic of passive behaviour) and at 60°C, the current increases were even more pronounced. A test carried out in a solution containing both gallic acid and sodium hypochlorite yielded very similar anodic polarisation behaviour to that obtained in the 50ppm gallic acid solution. This experiment with both gallic acid and sodium hypochlorite was undertaken in order to see if any possible reaction products (such as trihalomethanes), resulting from the organic acid and the chlorinated water, were likely to produce any substantial changes in the corrosion behaviour but, as Fig. 17 demonstrates, this did not appear to be the case.

3.7 CATHODIC POLARISATION IN TAP WATER

Trends observed in the more limited set of cathodic polarisation scans are shown in Fig. 18 which reveals extremely small cathodic currents in cold tap water and moderately heated water but higher currents at 60°C.

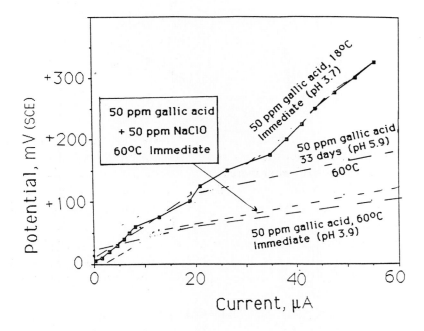

Fig. 17. Anodic polarisation BS 2871 copper in static water containing gallic acid.

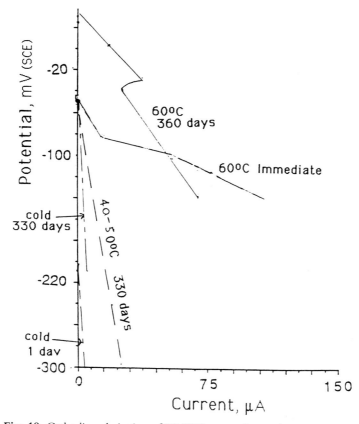

Fig. 18. Cathodic polarisation of BS 2871 copper in moving tap water.

3.8 pH CHANGES IN GLASGOW TAP WATER

Two series of tests were carried out in which the pH of tap water was measured over a period of time after drawing the sample from the tap.

In one test, samples of water were taken from the tap over a period of five days. A bottle and a beaker were partly filled with the tap water and the bottle was then stoppered whilst the sample in a beaker was left open to the atmosphere. The subsequent measurements (Table 1) showed that the pH of the water open to the atmosphere underwent a fall in pH after one day from the original values of around 8.5 to about 6.8 and thereafter remained fairly steady. The samples in the stoppered bottle exhibited slower reductions in pH but eventually reached similar values (below 7).

In the other experiment, samples were taken from the tap every day for 13d but, in this case, the water was run for several minutes and the bottle was filled to overflowing to remove all air from the bottle. The completely full bottles were then stoppered and and left for 5d before removing the stopper and immediately measuring the pH. The results are presented in Table 2 and show that the pH had fallen to below 8 (by as much as one pH unit in the most extreme case).

Table 1 pH measurements on tap water samples stored in partly-filled, stoppered bottle or in a beaker open to the atmosphere

Sample	Immediate pH	1 day	2 days	pH after 3 days	4 days	5 days
Sample taken Day 1						
Stoppered bottle	8.5	7.2	7.0	6.9	6.9	6.8
Open beaker	8.5	6.9	6.8	6.8	6.8	7.0
Sample taken Day 2						
Stoppered bottle	8.5	7.3	7.1	7.0	6.9	
Open beaker	8.5	6.8	6.8	6.8	6.9	
Sample taken Day 3						
Stoppered bottle	8.5	7.2	7.1	7.0		
Open beaker	8.5	6.8	6.9	6.8		
Sample taken Day 4						
Stoppered bottle	8.5	7.3	7.2			
Open beaker	8.5	6.9	6.8			
Sample taken Day 5						
Stoppered bottle	8.5	7.4				
Open beaker	8.5	6.9				

Table 2 pH measurements on tap water samples after storage in completely-filled, stoppered bottles for five days

Day of Sample	1	2	3	4	5	6	7	8	9	10	11	12	13
Immediate pH	8.6	8.5	8.2	8.5	8.1	7.9	8.2	8.0	8.4	8.1	8.2	8.4	8.6
pH after 5 days	7.6	7.8	7.6	7.7	7.7	7.7	7.9	8.0	7.9	7.8	7.6	7.7	8.0

4. DISCUSSION

4.1 Behaviour in tap water

The low currents recorded in the anodic polarisation tests upon commencing the potential shift positive to the natural free corrosion potential (E_{corr}) provided a demonstration of an inherent passivity exhibited by copper over long periods of time (up to one year) in cold water and moderately heated (40–50°C) tap water. At sufficiently positive potentials, the currents rose to higher values signifying breakdown of the protective films on the copper. It is not the intention herein to engage in a detailed analysis of these breakdown potentials and their characteristics but, rather, to emphasise the main consistent feature of the numerous anodic polarisation tests undertaken in this investigation, namely the passive behaviour over a significant range of potential positive to E_{corr}.

The values of E_{corr} were observed to move naturally in the positive direction. This aspect of the behaviour of copper might be considered to be potentially deleterious as it involves a move towards the positive values at which film breakdown occurs.[3,6] However, the major part of this natural drift occurred in the initial days of exposure and, as the anodic polarisation tests confirmed, this did not lead to breakdown of passivity.

The anodic polarisation tests carried out at 60–65°C indicated non-passive behaviour but examination of unpolarised specimens did not reveal evidence of corrosion except possible incipient pitting at the longest exposure of about one year. Moreover, the metal loss calculations were indicative of relatively low corrosion rates throughout the one-year period. Taken together with the observation of passivity upon immediate exposure of specimens to static tap water, these observations indicate that the copper is on the borderline between protective and non-protective behaviour at 60°C and above. The observed, longer-term, variations in E_{corr} (Fig. 11) might represent a further manifestation of this borderline behaviour since the excursions in the negative direction might have represented sequences of active corrosion followed by recovery of passivity.

The low currents recorded in the cathodic polarisation tests, even after extended exposure, indicated that copper does not present a very effective cathodic surface to support high current flow between anodic and cathodic areas in a macrocell once pits have initiated. However, the cathodic activity was seen to be greater at 60°C and, in any case, a low cathodic current density is counteracted by the area–ratio factor (i.e. high ratio of cathodic : anodic surface areas).

One interesting feature of the work involved the behaviour of samples taken from a pipe which had suffered severe pitting attack during service in a Glasgow hotel. From the results obtained with this material, it would appear that there was nothing uniquely poor about its corrosion resistance.

In summary, although seemingly marginal in corrosion resistance at 60°C, both grades of copper studied provided extensive evidence of corrosion resistance at lower temperatures. In contrast to these observations, pitting corrosion of copper has been found to occur in cold water service installations in Scotland as well as in hot water pipework. However, in practice, the onset of pitting in cold water systems appears to involve longer exposures than the one-year period employed in this study. This seems to point to the occurrence, in buildings supplied by Scottish public water, of certain environmental conditions (physical or chemical) that eventually become sufficiently aggressive to overcome the basic corrosion resistance exhibited by copper at these lower temperatures. These environmental factors may involve the settlement of water-borne deposits which are often observed in copper pipes removed from service in Scotland and were also noticeable in the current experimental work.

4.2 Influence of some water chemistry factors

Although the pH of the water freshly drawn from the tap was generally above 8, reductions were observed during subsequent storage. The observed pH drops, reproduced in Table 1, were initially thought to be simply due to absorption of atmospheric carbon dioxide followed by hydrolysis reaction to yield the weak acid, carbonic acid:

$$CO_2 + H_2O = H_2CO_3$$

On this basis, it could have been rationalised that the slower pH decline, in the stoppered bottles compared to that in the partly-filled container, was due to the more limited air supply. However, the results obtained from the bottles filled to overflowing (Table 2) indicate that additional factors can contribute to pH reductions which have been suggested[7] to be due to complexing reactions between humic acid and Ca, Al and Fe. In any event, the results quoted herein provide evidence that water left standing in copper pipework in a building could undergo reductions in pH.

The vulnerability of copper to corrosion, in low-pH water heavily loaded with organic acids (either natural or synthetic), was demonstrated in the anodic polarisation experiments in this study. Similarly, a highly-chlorinated water was shown to be corrosive to copper. Although utilising a chlorine concentration far in excess of the levels present in public supply water, this observation may have relevance to conditions arising as a result of the disinfection procedures employed during commissioning in large buildings. Other work[8] has indicated that the presence of chlorine, at concentrations considerably lower than those employed in the experiments described herein, may promote pitting in copper in the longer term.

5. CONCLUSIONS

1. In Glasgow tap water, copper has been observed to exhibit passive behaviour for extended periods of around one year in cold tap water and also at temperatures up to 50°C.
2. At temperatures of around 60°C, electrochemical tests have not revealed systematic evidence of similar passivity but have provided indications that the resistance of the material to significant corrosive attack may be inherently marginal at these higher temperatures.
3. Cathodic reactions appear to occur only at low rates — especially in cold tap water. This implies that the practical pitting phenomena involves macrocells with large cathode/anode area ratio.
4. The unbuffered nature of Glasgow water makes it prone to changes in pH in the acidic direction.
5. Exposure of copper to untreated, natural (loch) water or to gallic acid induces non-passive behaviour even at ambient temperature. This feature is associated with low-pH conditions.
6. Exposure of copper to high-chlorine environments induces non-passive behaviour.

6. ACKNOWLEDGEMENTS

One of the authors (J. A.) would like to express gratitude to the Pakistan Government for financial support for his studies at Glasgow University. The authors also make acknowledgement to A. Carver who carried out some of the experiments and to Prof. B. F. Scott, Head of Department of Mechanical Engineering, Glasgow University for the provision of experimental facilities.

REFERENCES

1. J. Nuttall, This Conf.
2. H. S. Campbell, Proc. 2nd Int. Congress on Metallic Corrosion, New York, 1963, 237.
3. V. F. Lucey, Proc. Int. Symp. on Corrosion of Copper and Copper Alloys in Buildings, Tokyo, 1982, 1.
4. J. C. McLuckie, Paper given to Scottish Branch, Inst. Hospital Engineers, Glasgow, Nov. 1992.
5. M. Woodward and T. Hodgkiess, unpublished work, University of Glasgow.
6. E. Mattsson, *Brit. Corros. J.*, 1978, 13, 5.
7. A. F. Walker, Ph.D Thesis, University of Strathclyde. 1985.
8. K. Kasahara et al., *Corrosion Eng. (Japan)*, 1988, 37, 361.

DISCUSSION

Mr G G Page began the discussion by reporting that similar buildings are affected by corrosion of copper pipework in New Zealand as in Glasgow. However, the problem there is one of 'blue water' rather than pitting. He emphasised the need to conduct tests under the same surface area/volume ratio conditions as in the practical application. Laboratory experiments sometimes use containers that are too small, leading to too high a surface area/volume ratio.

Mr R D Davies of R&D services asked if the aluminium sulphate treatment used to get rid of coloration from loch waters affects pitting. Dr Hodgkiess replied that alum additions and flocculation might be used elsewhere in Scotland to pre-treat water, but in the Glasgow area it is simply filtered. This tends to rule out a critical role for aluminium sulphate in teh pitting problem.

Mr A D Mercer of The Institute of Materials wondered what happened thirty years ago to prevent the widespread pitting now being encountered. Dr Hodgkiess did not know why the problems were now emerging. Corrosion is affecting pipes installed just two or three years ago in some cases but considerably older systems in others. He postulated that changes in water treatment might have occurred during the past decade adn been involved in the problem.

Dr J L Nuttall of IMI Yorkshire Copper Tube Company said that the observations that copper displays generally passive behaviour and that organisms play a role in the corrosion process could be important because the low flow rates encountered in large buildings might allow humic materials to deposit, thus explaining why the problems are concentrated in such establishments. However, Dr Hodgkiess reported that evidence is now emerging that small houses are also being affected.

Influence of Operating Conditions on Materials and Water Quality in Drinking Water Distribution Systems

I. WAGNER
DVGW, Karlsruhe, Germany

1. INTRODUCTION

Corrosion problems in water distribution systems are generally related to the materials used and their properties in combination with the quality of the water being distributed. To avoid or resolve adverse effects with regard to the corrosion behaviour of the systems, specific improvements in material quality and a qualified choice between the different available products should be made. The influence of the corrosion products, their solubility under different water quality conditions and, specifically, the uptake of metals into the water is a major concern. This is taken into account by the elimination of, or the restricted use of, materials with regard to the transported water quality.

As well as these direct material/water-related correlations, the operating conditions play an important role. Often, their influence dominates the behaviour of the systems and increases or decreases the corrosion processes with respect to the materials as well as to the water quality.

In this context, it can be seen that different operation conditions can lead to unexpected adverse effects on both materials and water quality.

2. RED WATER PROBLEMS

Iron and steel corrode in oxygenated water by the oxidation of the metal to ferrous ions and the production of OH^--ions, leading to depletion of the oxygen content in the water. Further oxidation of ferrous to ferric ions results in the production of rust, forming the well-known rust layers on the inner surfaces of the pipes.

The depletion of oxygen, caused by the corrosion process, usually does not affect the quality of the water because of the relatively small amount of oxygen reduction. But under unfavourable conditions, such as long transport times, high surface/volume ratios (small pipe diameters) or stagnation periods, the depletion of the oxygen concentration in the water can reach significant levels. In these circumstances water distribution systems often suffer from red water problems and this leads to consumer complaints.

Red water problems are often attributed to the mechanical mobilisation of precipitated rust slurry, but in fact these problems are mostly linked to the so-called 'non-stationary' corrosion process, non-stationary with regard to the change of the oxygen concentration during operation time.[1]

If the oxygen concentration in the area of a rust layer decreases too far, the corrosion process should come to a halt, because of the suppression of the cathodic part of the corrosion reaction. But certain components of the ferric corrosion products are able to substitute oxygen as the cathodic reactant, being reduced to the ferrous state and allowing the continuation of the corrosion by a process which can be described by the general equation

$$2\ Fe^{3+} + Fe \rightarrow 3\ Fe^{2+}$$

leading to a steep increase in soluble ferrous ions in the water.

Figure 1 shows this effect on a mild steel pipe with a preformed rust layer, where the straight line describes the maximum possible concentration of ferrous iron produced, depending on the oxygen concentration and the curved line shows the measured ferrous iron concentration in the water. The sharp increase in the range of low oxygen concentrations is directly related to the process described above.

Practical experience has shown that red water problems occurred with significantly higher probability in waters with increased neutral salt (sulphate, chloride) content, than in well-buffered waters with low concentrations of these anions.

The result of a study on these correlations is shown in Fig. 2, where the iron uptake rate was measured with two basic water qualities of different m-alkalinity and the addition of increasing concentrations of chlorides and sulphates.

The investigation has shown that a sharp increase of iron uptake rate occurs at neutral salt/alkalinity ratios between 0.7 and 1.

Finally, it was demonstrated that an important participant in the above mentioned reaction was the rust component γ-FeOOH (lepidocrocite), which can be reduced back to ferrous iron, and that a decrease in alkalinity and an increase in chlorides and sulphates promotes the production of this species, while in

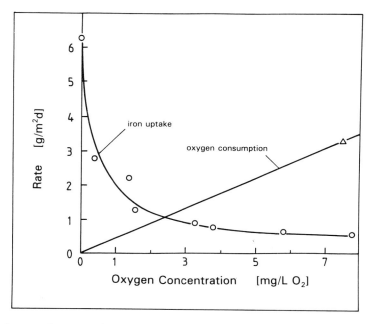

Fig. 1. Iron uptake rate (ordinate) vs oxygen concentration in the water in a mild steel pipe with inner rust layer.[2]

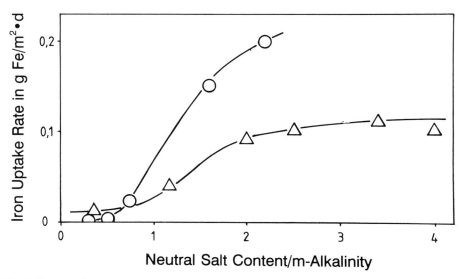

Fig. 2. Iron uptake rate of waters of different m-alkalinity and different neutral salt/alkalinity-ratios (in equivalents) measured in mild steel pipes which were under operation for 9 months in the different water qualities.[3]

○ m-alkalinity 5.5mmol L^{-1} △ m-alkalininty 1.5mmol L^{-1}.

opposite circumstances the α-FeOOH (goethite), which is stable and cannot become reduced from the ferric to the ferrous state[4] is the dominant species to be found.

3. USE OF INHIBITORS

To overcome corrosion problems, particularly the red water problems, the use of inhibitors (phosphates or phosphate/silicate-combinations) is a successful measure.

However, the expected or predicted success of such a treatment does not always occur because of the fact that operating conditions were not taken into account.

If an inhibitor is used, its efficiency depends on the ability of improving the quality of the corrosion product layers with regard to protective properties. This occurs with the consumption of inhibitor from the water into the corrosion products during the first stage of implementation.

This means that during transport, or more generally with increasing contact time, the inhibitor concentration in the water decreases, and pipe sections which are further away, may not receive sufficiently high concentrations during this phase. Therefore, it is necessary to start an inhibitor dosing programme at higher concentrations and reduce inhibitor concentration to the lowest possible values in a very careful manner. This effect of successive reduction of inhibitor concentrations with increasing operation time of the system was measured at the Duisburg Water Works, and is shown in Fig. 3.

The data show that any abrupt reduction of inhibitor concentration leads to an increase of iron in the water (for a limited period) until stabilisation occurs under the new conditions.

The case of unfavourable operating conditions which overcome the positive effect of inhibitor dosage is given in Fig. 4.

The curves show the influence of a central dosing with inhibitor on the iron concentrations at two different sampling points of the system. The curve marked 'Kellner Weg' represents the part of the distribution network with a more or less regular throughflow or, better still, a sufficient consumption of water to lead to a continuous fresh water input. The curve marked 'Sandweg' represents a part of the same system, where long transport time and low consumption at the end led to very low water exchange on that line.

The result of the consumption of the inhibitor and the insufficient feed of new inhibitor was a significant re-increase of red water formation and a failure of the inhibition for this part of the distribution network.

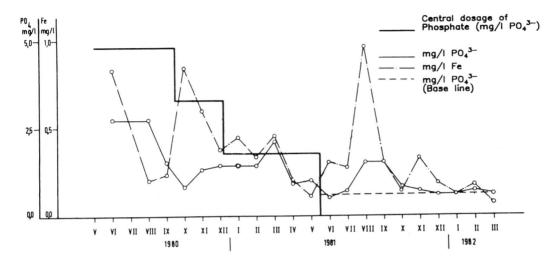

Fig. 3. Iron and phosphate concentrations at specific points of a water distribution network as a function of the original dosage of phosphates.[5]

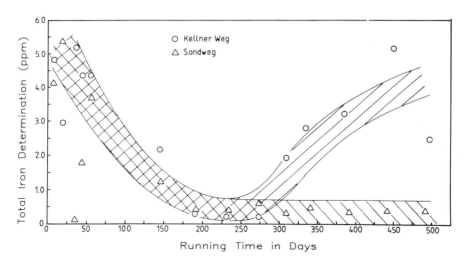

Fig. 4. Iron concentration in the water of a distribution network at two different sampling points after dosing with inhibitor.[6]

4. CEMENT-BASED MATERIALS

Asbestos cement pipes, cement mortar lining and concrete pipes are often under discussion with regard to their deterioration in calcium carbonate aggressive waters which dissolve the alkaline parts of the cements and lead to a destruction of the material. With regard to hygienic aspects, the possibility of an asbestos fibre release into the water is under discussion and is considered to be a deterioration of water quality or even a health risk.

In addition to these correlations between material properties and water quality, another effect takes place if soft and very soft waters are conveyed through pipelines made from cementitious materials. This effect is strongly related to the operating conditions and results in a pH-increase of the water.

As shown in Fig. 5, the pH of a soft water increases during the transport through a cement-mortar lined pipeline, to an extent which increases as the flow rate, expressed in terms of volume per day, decreases.

In this case, the basic pH of the water was in the range of pH 7 and increased at a flow rate of *ca.* $1500 m^3$ per day to a range of between 9 and 10. In other cases, where the original pH-value of the water is already in the range of pH 9, similar operation-related effects lead immediately to a water quality which does not comply with the Drinking Water Standards because of pH values over 9.5 and sometimes above pH 12, which corresponds to the pH of saturation with $Ca(OH)_2$.

The leaching of the alkaline parts of the cement not only results in an unacceptable water quality but also, in the long term, in continuous deterioration of the material similar to that caused by calcium carbonate-aggressive waters.

One way to overcome this kind of problem is to pretreat the cementitious surfaces before use. In the case of factory-made products, the exposure to wet air can improve the behaviour of the pipe with regard to the pH-problem, because a calcification (reaction of the $Ca(OH)_2$ with the CO_2 from the air to produce $CaCO_3$) of the surface takes place. This reduces or avoids the further transport of alkali into the water by clogging the surface and the pores of the material.

In the case of relining by cement mortar, the freshly rehabilitated (wet) pipeline can be purged by CO_2 gas, for example at a pressure of 1 or 2 bars, and the same calcification process takes place.

The kinetics of such a procedure were evaluated for a freshly cement-mortared piping system and Fig. 6 shows how the change of pH-development for a very soft water depends on the pretreatment procedure and the stagnation time after the water was introduced.

Influence of Operating Conditions in Drinking Water Distribution Systems 167

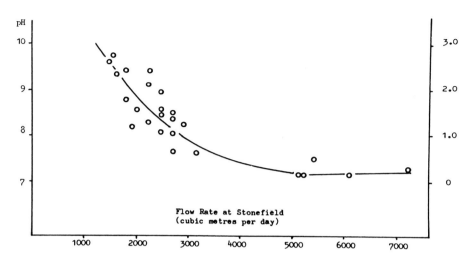

Fig. 5. pH-value of a drinking water after transport through a 9.4km long cement-mortar-lined ductile iron pipeline depending on the flow rate.[7]

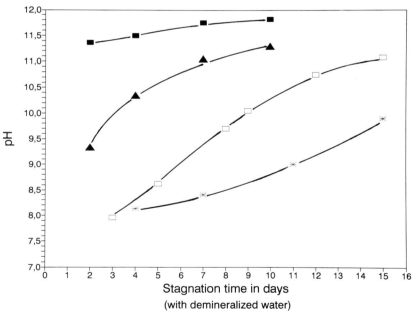

■ no pretreatment; ▲ 3 days, 1 bar CO_2; □ 7 days, 1 bar CO_2, 3 days, 2 bar CO_2; * 7 days, 1 bar CO_2, 6 days, 2 bar CO_2.

Fig. 6. pH-values of low-buffered water with stagnation time in cement-mortar-lined steel-pipes with different CO_2-treatment conditions.

5. METAL PIPES IN DOMESTIC INSTALLATIONS

Valuable information on correlations between materials used in distribution systems and the deterioration of drinking water exists for transport through domestic installations. In this context, the uptake of heavy metals during periods of stagnation is the most widely recognised example. In Figs 7 and 8 it is demonstrated by the uptake of lead into drinking water, that operating conditions mean more than stagnation and throughflow.

Figure 7 shows the calculated influence of stagnation time and pipe diameter on the lead content in the water after time t, $[Pb]^t$, in relation to the maximum concentration after lengthy stagnation $[Pb^\infty]$ and the initial concentration $[Pb]^o$. This graph shows for example, that in a 10mm pipe, 90% saturation is reached after only 2h of stagnation, while in a 50mm pipe the 2-hour-value reaches only 20% of the saturation value.

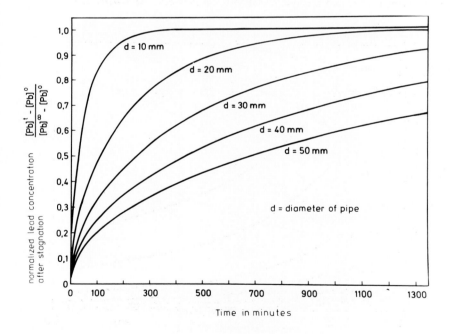

Fig. 7. Calculated relative lead concentrations in different pipe diameters vs stagnation time (calculated by mass transfer model).[8]

Equivalent data for the influence of flow conditions are shown in Fig. 8, where the relative lead concentration as a function of pipe diameter is given for different pipe lengths, and for different contact times during flushing.

As a last example, in galvanised steel pipes, the operating conditions can lead to an unexpected change in water quality. As with the corrosion of iron, oxygen depletion occurs as a result of the corrosion of zinc from galvanised steel pipes. Experience shows that in new pipes at pH-values below 7.3, the regular stagnation encountered in domestic installations leads to a serious decrease of oxygen concentration. In the case of low and very low oxygen concentrations, a nitrate-containing water provides nitrate as a substitute for oxygen. While the oxidation of zinc continues, nitrate becomes reduced cathodically to nitrite and subsequently to ammonia.

Because of the relatively very low permitted levels (MCLs) for nitrite and also for ammonia, these stagnant waters do not comply with the Drinking Water Standards.

Figure 9 shows schematically what happens as a result of this process during stagnation with decreasing concentration of nitrate, the intermediate role of nitrite, and, eventually, the almost complete conversion to ammonia.

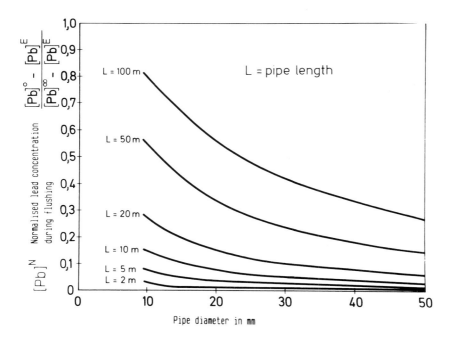

Fig. 8. Calculated relative lead concentrations for different pipe lengths vs pipe diameter (calculated by mass transfer model).[8]

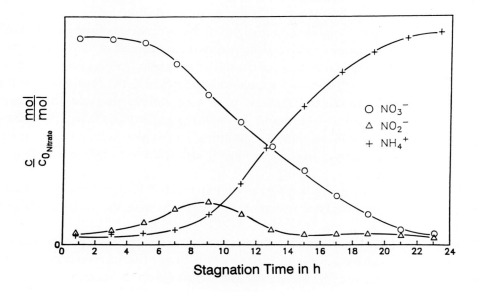

Fig. 9. Schematic variations of nitrate, nitrite and ammonia concentrations in a stagnant water in new galvanised steel pipes.

6. CONCLUSIONS

A survey of the different problems of corrosion damages of materials as well as deterioration of drinking water quality shows that a restricted view on the matter, taking into consideration material and water properties, cannot deal with the whole range of problems encountered. One of the main points to consider during evaluation of these complex correlations is the significant influence of operating conditions on most of the problems being encountered.

REFERENCES

1. A. Kuch, Investigations of the reduction and re-oxidation kinetics of iron (III) oxide scales formed in waters, *Corros. Sci.*, 1988, 28 (3), 221-231.
2. A. Kuch, Untersuchungen zum Mechanismus der Aufeisenung in Trinkwasserverteilungssystemen. Dissertation, Universität Karlsruhe, 1984.
3. I. Wagner, Die Bedeutung der Salzbelastung des Rheins für die Korrosionsprobleme in der Trinkwasserversorgung Tagungsbericht, 10. Arbeitstagung der IAWR 1985, 79 - 85. Hrsg. Sekretariat der IAWR, Amsterdam.

4. I. Dörfel, Untersuchung zum Verhalten reduzierter γ-FeOOH-Schichten in wässrigen Lösungen Studienarbeit, Fakultat fur Chemieingenieurwesen der Universitat Karlsruhe, 1988.
5. S. Felczykowski and I. Wagner, Die Umstellung der Wasserversorgung von Duisburg-Nord auf Trinkwasser aus Haltern, gwf-wasser/abwasser 1983, 124, (6), 289-295.
6. P. G. Schumacher, I. Wagner and A. Kuch, Die Trinkwasserversorgung von Göttingen mit Mischwasser - Erfahrungen über den Einfluß der Wasserqualität und von Inhibitoren auf Korrosion im Rohrnetz, gwf wasser/abwasser 1988, 129, 146-152.
7. H. H. Collins and T. R. Smith, Bore coatings for spun iron pipes. Water Distribution Systems, Paper 11. Papers & Proceedings, WRC-Conference, Oxford, Sept. 1978, Ed. WRC, Medmenham, Feb. 1979.
8. A. Kuch and I. Wagner, Transfer model to describe lead concentrations in drinking water, *Water Research* 1983, 17 (10), 1303-1307.

Assessment of Corrosion Trends in a Potable Water Distribution System Including an Automatic Analysis Technique

A. M. DEAN AND T. KOCH*

School of Engineering, Staffordshire University, Beaconside, Stafford, UK

1. INTRODUCTION

An initial project was carried out on potable cast iron water mains in the Stoke-on-Trent, Moorlands and Stafford areas of Severn Trent Water Ltd (STW) by Staffordshire University in 1984.[1] The objective of this project was to assess the corrosion resistance, strength reduction and flow resistance of the sample mains and approximately six hundred samples were examined. The corrosion resistance aspect of the survey was extended in 1985 to cover the whole of the Severn Trent area. To date approximately four thousand samples have been examined.

Individual samples are tested to assess the extent of external and internal corrosion and the amount of tuberculation present in the sample, the techniques used being similar to those used by other workers in the field.[2] The samples are not wholly random, some samples being opportunistic in nature.

The survey provides data on a number of different levels. Detailed information is available concerning individual pipe sample mains with a report reviewing the corrosion of a number of samples in a particular area. In addition to this, the database of sample results can be used to examine any trends in corrosion behaviour, which may be useful to Severn Trent Water for asset management planning.

The University realises the limitations of traditional analysis of corrosion data and has collaborated with Krupp Forschunginstitut GmbH, to use an automatic learning package to assess the data. Initial results from the analysis are included in

---

*Krupp Forschunginstitut GmbH, Informationstechnik, Muenchener Str. 100, W-4300 Essen 1, Germany.

Assessment of Corrosion Trends in a Potable Water Distribution System

this chapter and these indicate that the rules generated can be of use to engineers in assessing rehabilitation priorities within the distribution system. Additionally, alternative techniques for data display using a geographical information system are being examined and evaluated.

2. SAMPLE TESTING AND ANALYSIS OF CORROSION DATA

This paper has been restricted to the consideration of the internal corrosion and related factors, although both external and internal corrosion were examined in the survey. Results from the analysis of external corrosion are included only when a comparison value is needed or where the techniques being used apply to either corrosion type.

2.1 Information obtained during testing

The experimental values relating to internal corrosion measured during testing are detailed in Table 1. In addition to these measurements, data supplied by Severn Trent Water Ltd are shown in Table 2 and calculated values relating to internal corrosion are given in Table 3.

2.2 Analysis of the Corrosion Database

The results of the testing procedure conducted by Staffordshire University are entered into an Oracle database. When the project was started in 1984 commercial databases were not readily available for storage and analysis of specialist data and an in-house database was written by the Computer Services Department within the University. This database was used, with several modifications, until 1991 when the data were transferred to the new system. The original database was unwieldy and only a limited amount of analysis was possible. However, the Oracle database now in operation can be used in combination with a large number of packages for analysis purposes.

When the project was initiated Severn Trent Water Ltd was organised into eight Divisions. A serial numbering system based on a five digit code was developed so that each sample could be related to the Division of origin and water source. However, the company is now organised into fourteen Management Districts and a revised, seven digit coding system, is employed to relate a particular sample to the appropriate Management District and District Metered Area (DMA). A DMA is the smallest information collecting unit for management purposes.

Table 1 Measurements taken on each sample relating to internal corrosion

1. MEASURED INTERNAL DIAMETER: Average measured internal diameter of the pipe; based on four equally spaced measurements (mm).

2. AVERAGE WALL THICKNESS: Average original wall thickness based on eight equally spaced measurements (mm).

3. INTERNAL PIT DEPTH: Maximum internal pit depth. This figure represents the maximum internal pit depth measured from 10 sample measurements (mm).

4. AVERAGE TUBERCULATION HEIGHT: Average tuberculation height; based on 16 readings (mm).

5. THINNEST WALL REMAINING: Thinnest remaining wall thickness measured (mm).

Table 2 Data supplied by Severn Trent Water Ltd

1. MATERIAL: Material type.

2. AGE: Age of the pipe supplied by Severn Trent Water Ltd.

3. SOIL: Soil type surrounding the pipe; based on a qualitative assessment of the soil type.

4. WATER: Water type supplied by Severn Trent Water and based on analysis of sources.

Table 3 Calculated values relating to internal corrosion

1. INTERNAL CORROSION RATE: Internal corrosion rate. The maximum internal corrosion rate, calculated from the age and maximum pit depth, assuming a linear relationship.

2. CHAMPION CLASS INTERNAL: Champion classification* of width to depth ratio based on the average of 10 measurements.

3. TUBERCULATION PER YEAR: Average rate of tuberculation growth per year; based on the value of average tuberculation height and age and assuming that tuberculation growth is a linear function with time.

4. PERCENTAGE BLOCKED: Percentage of the pipe blocked, calculated on the assumption that the tuberculation can be represented by the average tuberculation height at all points around the pipe.

---

*From F. A. Champion, Corrosion Testing Procedures, 1964. Chapman & Hall, London, UK.

Initially, it was recognised by the University[1] that assessment of individual samples could provide only a limited amount of information and the corrosion data obtained may not be representative of the main from which they have been extracted. In addition to this, some of the data supplied by Severn Trent Water Ltd were subjective in nature. However, it was felt that analysis could be useful to the company assuming that the limitations were appreciated and that the results were used in conjunction with information from other sources within the company. It is important for any analysis to be related to the various operational Management Districts.

In the pipe samples extracted, internal corrosion could be related to a number of measured variables within the survey. Water type was an obvious variable but reservations as to the consistency of the supply were expressed by Severn Trent Water Ltd. It was decided that five qualitative water types should be identified to overcome this problem and these are detailed in Table 4. Material type was also seen as a variable in the corrosion process and the three types encountered in the survey are shown in Table 5.

A decrease in corrosion rate with increasing pipe diameter has been observed

Table 4 Water types identified in the survey

Type Number	Description
1	Upland Reservoir
2	Upland River
3	Lowland River
4	Borehole
5	Bulk Supply from other sources

Table 5 Material types identified in the survey

Type Number	Description
1	Vertically Cast Iron
2	Spun Iron
3	Ductile Iron

Table 6 Diameter ranges

Diameter ranges (mm)
< 80
80–110
> 110

for soft waters[3] in small diameter ferrous mains and three diameter ranges were identified for study. These are shown as Table 6. Age was included as a variable, although it is noted that in cast iron mains, age may not be particularly useful in identifying the extent of corrosion.

Traditional analysis of the data has been attempted by the University. In 1990 an interim report[4] detailed the results of the analysis of 185 random sample mains from the Southern Division of Severn Trent Water Ltd. Both external and internal corrosion were assessed and the analysis indicated that, although the number of samples from one area appeared large when the number of variables was taken into account, the size of the sample was too low for valid statistically correct conclusions to be drawn.

In 1992 the analysis was extended to include all the Management Districts of Severn Trent Water Ltd, but to date only a limited amount of useful information has been achieved because of the complex nature of these data. The difficulties of attempting such a programme should not be underestimated. The approach of a factorial analysis, where the influence of various environmental and pipe factors could be evaluated, would have been difficult to achieve. In this kind of analysis, the factors that influence the condition of a pipe are divided into levels. For the survey there were already fourteen different locations, nine types of soil, five levels of water, three levels of material type and three levels of pipe diameters. By including the age of the pipe with five different levels there are six influencing parameters with 28 350 possible combinations. Although it is likely that all cells would not exist for all management districts, the number of combinations would have been large. In addition the inclusion of opportunistic samples in the survey could bias the data and swamp any effects from individual levels. As the data is to be analysed within the Management District areas of Severn Trent Water to be useful for investment policy, it was decided not to carry out a factorial analysis as the initial strategy.

As a first stage in the analysis a number of 'gut' feelings were tested. The University was asked to investigate several relationships, including those between age and internal corrosion and age and tuberculation. Samples were to be sorted so that variations due to material type, water type and diameter could be investigated. This type of analysis, although not very sophisticated, does give some indication of the corrosion trends within the data and would enable confidence to be gained if the results happened to agree with previous surveys.

The relationship between pit depth and age is used to determine corrosion rates and life expectancies in samples. The generalised equation for corrosion in ductile and grey cast iron is of the form:

$$p = kt^n \qquad (1)$$

where p is the pit depth in time t; k and n are constants, n being generally less than 1.[5] It has been found from field trials[6] that n usually lies between 0.2 and 0.8, with a mean value of 0.44, the values being related to the corrosion process and the influencing variables.

As the values of k and n are unknown for an individual sample, a linear corrosion rate is usually used as this represents the 'worst case scenario'. If internal corrosion is considered, a linear corrosion rate may be more appropriate than the parabolic form of eqn (1). The mechanisms involved in internal corrosion, due to the growth of tuberculation in the inner surface of the pipe, are different to those generating external corrosion. The layer of corrosion product gradually stifles the corrosion reaction and a further decrease in the rate of corrosion may be observed.

Graphs of maximum internal pit depth vs age were plotted for each Management District, with material type being differentiated in one set, and water type being differentiated in the other. Examples of two of these graphs are shown in Figs 1 and 2. All the graphs were very similar and showed that the variation in maximum pit depth between individual samples of the same age increased up to approximately 60 years, then either remained constant or decreased. At the same time the value of maximum internal pit depth increased up until approximately 60 years, then remained constant or decreased. This indicates that the linear model for internal corrosion may be unrealistic for the older vertically cast pipes. As this rate of corrosion is often used to predict a life expectancy for the sample main, a model which more accurately predicts the corrosion mechanisms is required. It is particularly important when the lifetime prediction is within the planning horizon used.

No difference could be observed between samples supplied by different water types within the same Management District. It is likely that this arises because the supply might not have been constant throughout the life of the pipe.

The percentage of the cross-sectional area of the pipe which is blocked by tuberculation was plotted as a scatter graph for all samples within a water type for Severn Trent Water Ltd, Management District not being differentiated. Graphs for upland reservoir water are shown in Figs 3 to 6 (pp.180–183). A clear dependence on pipe diameter is seen in these graphs, the maximum proportion of the cross-section blocked seen in samples over 110mm dia. at any age being 50%. The results were repeated for all water types. This information could be useful in planning relining activity.

Mean maximum internal corrosion rates were determined for all Management Districts within water type and material type. The significance of average corrosion rates is doubtful, as pit depths do not form a normal distribution, but they do provide a quick and easy method of differentiating between the corrosion behaviour in different Management Districts.

The results of this preliminary analysis carried out by Staffordshire University

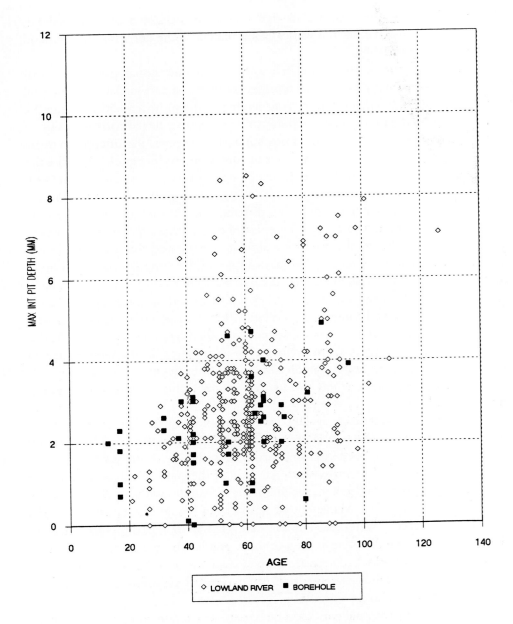

Fig. 1. Maximum internal pit depth vs age for Leicester.

may be viewed with caution and cannot be statistically proved. However, in general they agree with results from previous surveys within the water industry and this gives some confidence that future analysis may provide data of significant use, in relining and replacement programmes.

Assessment of Corrosion Trends in a Potable Water Distribution System

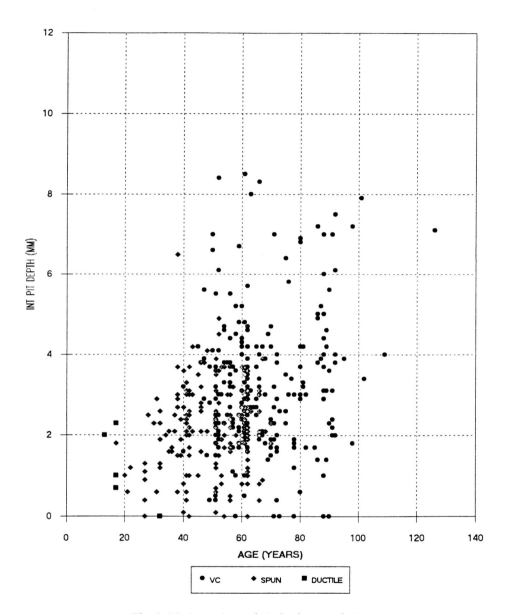

Fig. 2. Maximum internal pit depth vs age for Leicester.

3. THE RULEARN® AUTOMATIC LEARNING PACKAGE

The data contained in a database can often only be determined by the domain expert, on the basis of experience which allows any rules or relationships to be

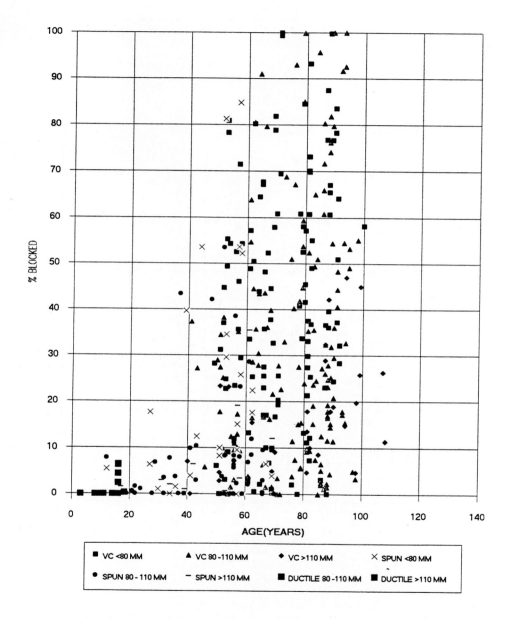

Fig. 3. Percent blocked vs age for all samples supplied by upland reservoir water.

recognised easily. However, if the data are large or complex, or there are a number of interrelating factors, even the domain expert can find it difficult to find a relationship. If the process changes from one state to another, the problem is even

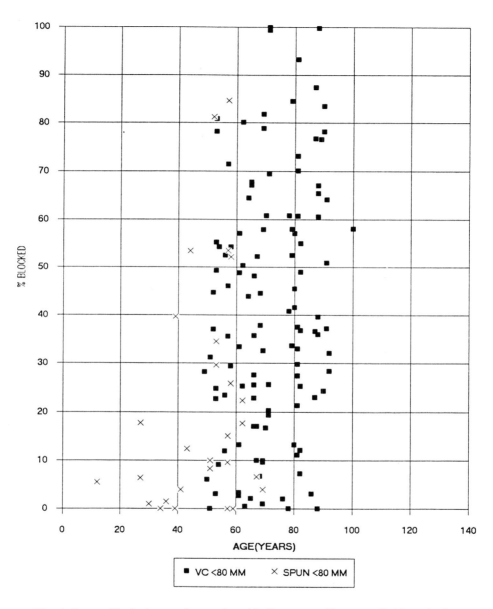

Fig. 4. Percent blocked vs age for samples with diameters < 80mm supplied by upland reservoir water.

more difficult to solve and in cases such as this, computer assistance in interpreting the data is desirable, if not essential.

RULEARN has been developed by Krupp Forschungsinstitut GmbH[7] as a tool

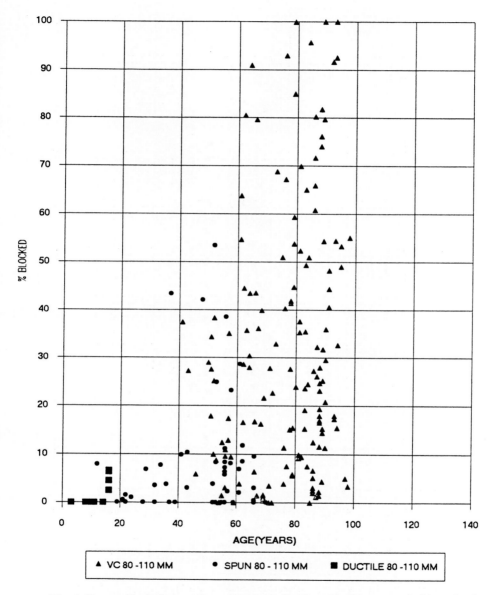

Fig. 5. Percent blocked vs age for samples with diameters 80–100mm supplied by upland reservoir water.

for automatically interpreting factual data. RULEARN learns the unknown dependencies underlying the facts in the database and makes them available to the specialist as rules.

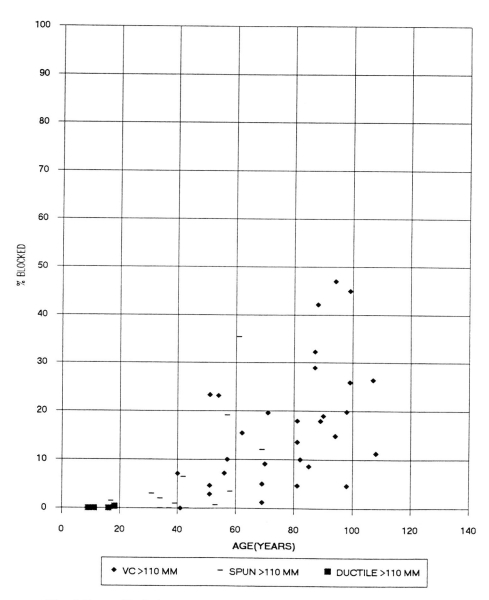

Fig. 6. Percent blocked vs age for samples with diameters > 110mm supplied by upland reservoir water.

The RULEARN module learns experience in the form of rules. The information made available to the learning process from outside stems from a fact basis.[8] RULEARN has been applied to data with more than one hundred influencing variables.

RULEARN learns rules which exhibit sufficiently high quality. The quality of the rule is described by three values — reliability, specialisation and universality:

$$\text{Reliability} = \frac{\text{Number of correct applications of the rule}}{\text{Number of possible applications of the rule}}$$

$$\text{Specialisation} = \frac{\text{Number of tests for which the target variable is true}}{\text{Number of tests in which the rule is applicable}}$$

$$\text{Universality} = \frac{\text{Number of cases in which the rule is applicable}}{\text{Number of tests}}$$

Before the learning process begins the user specifies the minimum reliability, the minimum universality and the maximum specialization. This requires some feel for the quality of the database and the effect of the control parameters.

The analysis that RULEARN can provide offers advantages such as data compression and the use of both numerical values and symbols in the source data. RULEARN can also generate rules if the data are incomplete or uncertain and there is no limit to the number of parameters influencing the target variable. The quality of the results can be adjusted by means of the control parameters and the system adapted to different databases.

3.1 The RULEARN analysis

A raw set of data from approximately 4000 pipe samples was supplied to Krupp Forschungsinstitut GmbH for analysis with the RULEARN package. Samples, which were known to contain corrupted or unreliable data, were removed leaving 3810 samples for analysis.

The main target variables highlighted by the University were internal and external pit depths, corrosion rates, percentage of cross-section blocked, tuberculation growth rate per year and life expectancies. The influencing variables were location, material type, age, diameter and soil and water types.

Rules were extracted for the target variables and given a value of reliability (R) for the number of cases described (N). RULEARN found a variety of rules. A typical example of a rule generation table can be seen as Table 7. Useful rules may be extracted from the Table, e.g. Rule 12 for internal corrosion indicates that in the Coventry and North Warwickshire Management District (Loc 09), in pipes of less than 110mm i. d. (Dia <= 110.00) supplied by lowland river water (Wat 3), the internal corrosion rate was more than 0.03mm/year (=> In Cor > 0.038). The value of R in this case was 0.742 and that of N was 236.

Table 7 A rule generation table

Rules for Internal Corrosion Rate

	R	N								
Rule 1	0.783	46	Loc_12	&	Dia>110.000			=>	In_Cor<=0.038
Rule 2	0.755	49	Loc_11	&	Dia>110.000	&	Mat_2	=>	In_Cor<=0.038
Rule 3	0.786	56	Loc_11	&	Dia>110.000	&	Wat_3	=>	In_Cor<=0.038
Rule 4	0.741	112	Loc_11	&	Mat_2	&	Soil_1	=>	In_Cor<=0.038
Rule 5	0.756	45	Loc_15	&	Dia<=110.000	&	Soil_4	=>	In_Cor<=0.038
Rule 6	0.754	65	Soil_2	&	Dia>110.000	&	Wat_4	=>	In_Cor<=0.038
Rule 7	0.867	45	Loc_01	&	Soil_2			=>	In_Cor>0.038
Rule 8	0.773	44	Loc_02	&	Wat_4			=>	In_Cor>0.038
Rule 9	0.762	42	Loc_07	&	Mat_1			=>	In_Cor>0.038
Rule 10	0.779	95	Loc_08	&	Mat_1			=>	In_Cor>0.038
Rule 11	0.758	132	Loc_09					=>	In_Cor>0.038
Rule 12	0.742	236	Loc_09	&	Dia<=110.000	&	Wat_3	=>	In_Cor>0.038

R = Reliability of the rule
N = Number of cases described

Rules from different generation tables can be combined to give additional information concerning a particular Management District. In addition to Rule 12 above, RULEARN revealed in another table that, in Coventry and North Warwickshire, the tuberculation rate was greater than 0.075mm/year in vertically cast samples of less than 110mm internal diameter and that the internal corrosion rate was more than 0.038mm/year for all vertically cast samples. As well as information concerning internal corrosion, RULEARN detected that in sand/gravel soils in Coventry and North Warwickshire the external corrosion rate was less than 0.038mm/year.

Such simple rules are useful to the various engineers within a water company. For engineers working within a Management District the rules can confirm experience in the field and suggest areas of future study. For rehabilitation engineers different Management Districts can be compared and evaluated with regard to their corrosion history and the rules can be used in conjunction with other information concerning the distribution system, to assist in making decisions regarding replacement or relining.

The learned rules confirm previous experience within Management Districts. If the target variables and influencing parameters had been modified, even more detailed rules would have been generated.

4. THE ARC/INFO GEOGRAPHIC INFORMATION SYSTEM

Many of the tasks undertaken by the rehabilitation departments of water companies are dependant upon geographic relationships. A Geographical Information System (GIS) is a computer system designed to manage spatially referenced data and to provide users with the tools to answer those questions based on geographical relationships.

The ARC/INFO system integrates maps and attributes information in a common spatial database. Powerful data management, manipulation and modelling tools allow data to be retrieved rapidly, perform sophisticated analysis and output the results as tabular reports or high quality maps. The system integrates advanced cartographic facilities with a relational database management system (RDMS) to provide the capabilities for automation, management analysis and display of spatial information.

The geographical area covered by Severn Trent Water Ltd is divided into fourteen Management Districts and within each district the smallest identified unit is the District Metered Area (DMA). Each Management District has been digitized by the University and the DMA outlines identified. Where Management Districts included a centre of urbanisation the concentration of DMAs increased and for clarity these areas have been digitized separately but linked to the main digitised maps for data display. An example of a digitised Management District showing DMA numbers is shown in Fig. 7.

Information which is of use to the rehabilitation engineer can be related to a unique DMA number within a Management District. Parameters that could be of interest are discolouration, taste and odour, chlorinous taste, iron customer tap failures, free chlorine residuals and related customer complaints. All of these parameters can be coded according to either externally or internally determined criteria and the data displayed within the relevant Management District of the company according to the DMA reference. An example of a map produced to specific criteria is shown as Fig. 8. The ease of interpretation of information supplied by such a system is evident as areas of concern are immediately apparent.

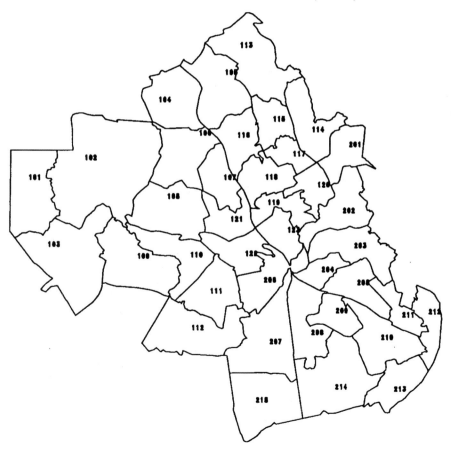

Stoke District (2). District Meter Areas (DMAs).

Fig. 7. Digitised management district showing DMAs by number.

Fig. 8. WSIR results 1991.

Stoke District (2). District Meter Areas.

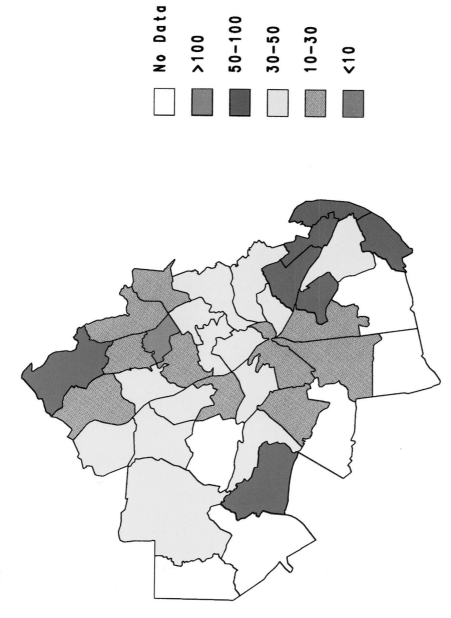

Fig. 9. Staffordshire University corrosion data — total corrosion priority life.

In addition to the data display provided in Fig. 8 information contained in the corrosion database may be similarly displayed. Life expectancies based on total corrosion, external corrosion and internal corrosion could be used and these related to DMA number. Examples of maps plotted to such criteria are shown in Figs 9 (previous page), 10 and 11(shown on pp.192 and 193). Areas of greatest tuberculation could also be highlighted with ease within a Management District.

The information that could be displayed on such a system has great variety. A display map can be viewed across a continuous map area and the level of detail controlled.

Relationships between interrelating factors can be investigated by the use of map overlay. Map overlay enables two or more maps and their attribute files to be combined and a new map created. The system establishes the spatial relationships of the combined map features and associated attribute data from the original maps. At present the information supplied has been related to DMAs only, but other geographical information could be added, for example, the position of the trunk mains within the Management District could be digitised or soil survey information included.

The ARC/INFO system also allows for the continual updating of the digitized maps as boundaries of DMAs change or combine so that 'work done' is not lost as on a traditional paper system. Data transfer is also electronic, ARC/INFO providing an explicit interface with other RDMS with modest custom programming investments. The interface will allow Standard Query Language (SQL) queries from database management systems such as Oracle. This open architecture solution encourages better data management and gives users greater freedom to integrate a variety of data management technologies.

5. CONCLUSIONS

The limitations of the statistical analysis of the corrosion database related to water mains operated by Severn Trent Water Ltd have been demonstrated by Staffordshire University and although some of the results of such an analysis may confirm experience in the field, it is difficult to justify investment using these results as the levels of confidence experienced are low.

The RULEARN automatic learning package has been shown to be a suitable tool for evaluating a database containing corrosion information from a distribution system. The rules extracted from the RULEARN analysis could be used in conjunction with other data indicators concerning the distribution system to assess the investment strategy of the water company.

The display of data indicators mentioned above can be achieved easily with the use of a Graphical Information System package such as the ARC/INFO system at Staffordshire University. The data can be continually updated and a variety of overlay techniques can be used to investigate inter-dependencies.

The rehabilitation of water mains is becoming increasingly important as distribution systems age. It is necessary for water companies to be aware of the variety of techniques that are available for the analysis and display of data to enable them to make informed decisions concerning their investment strategy. Several techniques have been presented in this paper and their relevance to the water industry has been assessed.

6. ACKNOWLEDGEMENTS

The authors would like to acknowledge the assistance of Severn Trent Water Ltd in the preparation of this paper.

REFERENCES

1. A. M. Dean, H. C. Lowe and A. P. Parker, An Investigation into the Corrosion, Strength Reduction and Flow Resistance of Cast Iron Water Mains in the Potteries, Moorlands and Stafford Areas of the Severn Trent Water Authority, Stafford, 1984, North Staffordshire Polytechnic.
2. M. Randall-Smith, A. Russell and R. Oliphant, Guidance Manual for the Structural Condition Assessment of Trunk Mains, Swindon, 1992, Water Research Centre.
3. S. Williams, R. G. Ainsworth and A. S. Elvidge, A Method of Assessing the Corrosivity of Waters towards Iron, WRc, Technical Report, 1984.
4. A. M. Dean, A. Hughes and C. J. Kitchen, Preliminary Statistical Investigation of Pipe Sample Results for Severn Trent Water Plc, Stafford, 1990, Staffordshire Polytechnic.
5. R. A. Gumnow, *Material Performance*, 1984, 23, 39-42.
6. E. C. Sears, *Material Protection*, 1968, 7, 33-36.
7. S. Harris, B. Fehsenfeld, T. Koch and R. Kirchheiner, *Techn. Mitt. Krupp*, 1991, 40, 27-36.
8. T. Koch, Inductive Learning of Compact Rule Sets by Using Efficient Hypotheses Reduction, TR 92-069, Berkley, CA, 1992, International Computer Science Institute.

DISCUSSION

Mr Mercer asked whether the type of analysis that Ms Dean had described would be extended from the main distribution network to include the water supplies to houses. Ms Dean replied that this was not planned but that there would be nothing to stop the approach being used flexibly for such purposes.

(Discussion continued on p.194)

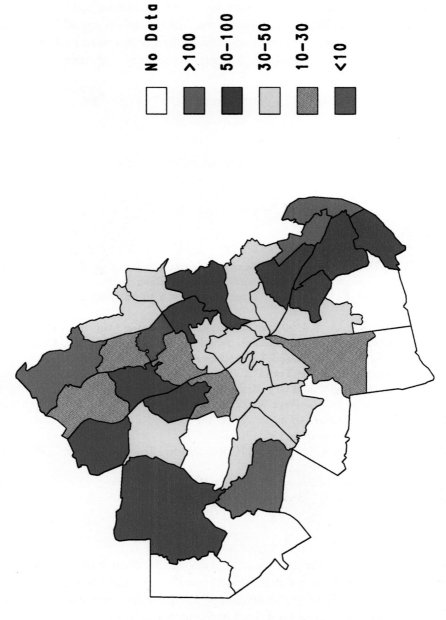

Fig. 10. Staffordshire University corrosion data — internal corrosion priority life.

Stoke District (2). District Meter Areas.

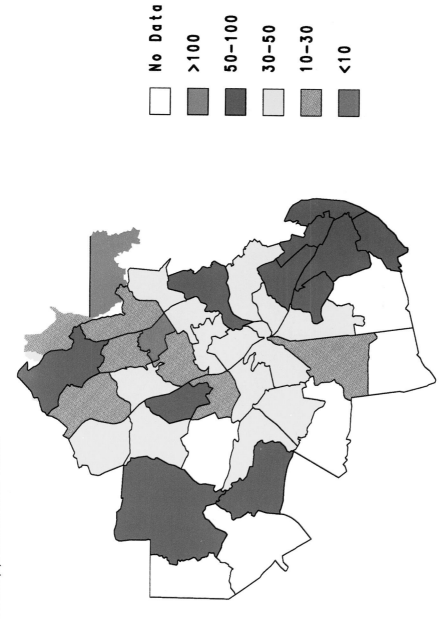

Fig. 11. Staffordshire University corrosion data — external corrosion priority life.

Prof. Fischer commented that while RULEARN is good for dealing with historical data stored in databases, the value of the output depends upon the stored data and new information cannot thereby be created. Therefore, he stressed the need for care to ensure that effort was not diverted from research to provide better information about mechanisms, etc.

Influence of Network Disinfection on Corrosion of Galvanised Steel Pipes (Use of Hypochlorite, Permanganate and Hydrogen Peroxide) — Case of Paris

P. LEROY

Centre de Recherche et de Contrôle des Eaux de Paris,
156 Avenue P. Vaillant-Couturier, F-75014, Paris, France

ABSTRACT

Generally a disinfection of the household potable water network is needed before connecting to the public network. Such an operation is actually carried out in all buildings built in Paris and its suburbs. The chemical used to date is potassium permanganate. This choice results from the high coloration of the product, which simplifies the operation performed by the plumber. However, numerous cases of corrosion take place as a result of this preliminary treatment. A laboratory experiment, lasting about one year, shows the influence of the nature of the product on the behaviour of the galvanised steel pipes. It appears that manganese dioxide deposits formed on the inner pipe walls induced corrosion.

On the other hand, the use of hydrogen peroxide lead to a rapid attack of the layer of zinc initiating the formation of protective deposits of zinc products. In such cases the behaviour of the pipes is very favourable.

The use of hypochlorites or chlorine does not affect the pipe wall and has little or no effect on the corrosion of the pipe.

1. INTRODUCTION

The water supply system of Paris was designed during the last century by the engineer Eugène Belgrand in the middle of the last century. He decided to supply Paris by spring water drawn 100km away from the city in the West and in the South. The increasing water requirement from year to year has made it necessary to use treated surface water taken from the Seine and Marne rivers.

At present, Paris is supplied by two types of water:

- spring or underground waters drawn *ca.* 100km from the west and the south of the city, and transported in aqueducts by gravity without any other energy. The spring waters are drawn from karstic areas in the chalk. The underground water is drawn by drillings in the alluvium water tables of river Seine or some other influents of Seine (Avre, Yonne, etc.).

- surface water taken from the river Seine and river Marne, treated in three different plants. Two plants work with a technique of slow sand filtration (St. Maur and Ivry) and one with a technique of rapid filtration (Orly).

These two types of resources produce, in approximately equal parts, the 800 000m^3 necessary to feed Paris daily.

Spring and underground waters directly feed several tanks inside Paris from which water is distributed by gravity to the consumers. These tanks mainly supply the lower parts of the city in the western part and along the river Seine. Waters from the treatment plants are pumped to a pressure of approximately 10 bars and supply the upper parts of Paris, the eastern part and peripheral southern and northern parts as indicated in Fig. 1.

Zone A corresponds to the southern spring water, zone B to river Seine treated water, zone C to western spring water and zone D to Marne treated water. Zones marked A+B are fed by a blending of the two types of water.

Both kinds of water are hard waters of between 250 and 300ppm $CaCO_3$. As it comes from springs or rivers, the water introduced into the distribution network presents a colour less than 5 units and a turbidity that exceeds 0.1 NTU for only a few days during winter swelling. They are either naturally calcifying or treated to be slightly calcifying before introduction in the distribution network.

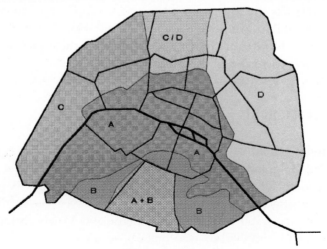

Fig. 1. Origins of distributed water in Paris.

So, for more than a century, the inhabitants of Paris received sufficient daily water of good quality. It is perhaps the reason why they are so exigent with the quality of the water coming from their taps. The head of the municipalities is well aware of this Parisian characteristic and is very heedful to all water problems.

In striving to satisfy the consumers the water service of Paris receives and studies all the complaints of the inhabitants. It appears that according to the physico–chemical properties of the distributed water the quality of the water is rarely significantly degraded in the public network. Paris does not suffer from 'red water problems'. But the main source of quality degradation is in the household network, and principally the corrosion of galvanised steel pipes. Figure 2 gives the main grounds of complaints and shows the importance of colour and aspects problems.

In such conditions it is not possible to say that the corrosion of the household network is due to the aggressively of the distributed water. It appears that most cases of corrosion are due to the design and the realisation of the network. But in some cases, in new, large networks not more than ten years old, important corrosion appears without any clear origin.

On the other hand, in Paris, all new networks have to be disinfected before buildings are opened to the users. The question is to discover if such treatment can have an effect on the behaviour of the network.

2. DISINFECTION PROCESS

In France disinfection of the household networks of collective buildings is required before allowing the inhabitants to move in. After hydraulic tests of the network, a powerful flushing operation is carried out to remove metal filings, excess grease or other impurities. Then, a disinfecting agent is introduced in the water entering the network near the water-meter. The reagent used is either sodium or calcium hypochlorite, or potassium permanganate.

The injection conditions of the reagent are such that the disinfecting concentration of the water in all the parts of the network is:

–100mg L^{-1} of free chlorine when hypochlorites are used
–150mg L^{-1} of potassium permanganate if this reagent is used.

Contact between these solutions and the network is maintained for *ca*. 24h after which a new flushing is carried out to purge the network. When all the disinfecting solution is eliminated, water stays in the network for a further 24h before taking samples of water for bacteriological examination.

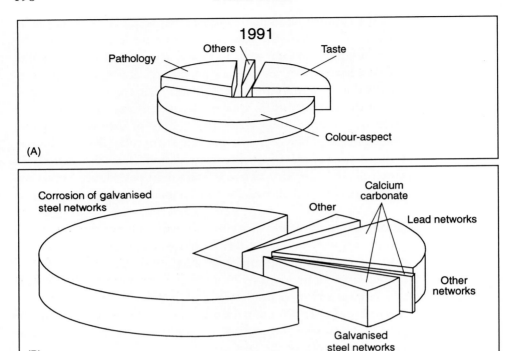

Fig. 2. Main grounds of complaints (A) and origins of aspect degradation by deposits found in tap water (B).

Potassium permanganate is generally used in France for this operation. This choice results from the fact that this reagent is highly coloured and can be detected visually by non-chemists, the disinfection being carried out by plumbers.

3. MATERIALS AND METHODS

Laboratory experiments have been performed in a way to simulate disinfection treatment on different elements of a small sized network. In this experiment we compared the effects of the treatment using three different reagents: sodium hypochlorite, potassium permanganate and hydrogen peroxide.

3.1 Laboratory setting used

Three elements, similar to the one shown on Fig. 3, has been used. They are composed of horizontal and vertical parts made of galvanised steel and copper. Threaded and brazed fittings are used to show the effect of disinfection on specially sensitive parts of networks.

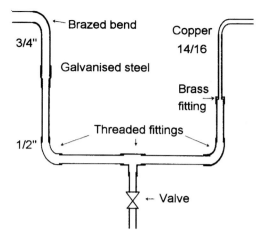

Fig. 3. Laboratory setting used.

The three elements of the network have been first polluted by passing raw river water through them for 10h. Then they have been flushed and disinfected with the reagents indicated previously. The concentrations used are as follows:

- NaClO, 100mg L^{-1} of chlorine;
- KMnO$_4$, 150mg L^{-1} of reagent;
- H$_2$O$_2$, 1000mg L^{-1}.

Disinfection of the three elements was conducted as indicated above for private networks.

3.2 PROCEDURE OF THE EXPERIMENT

After disinfecting, a first series of pipe samples has been taken on the horizontal parts of the galvanised steel pipes near threaded fittings in order to make a preliminary observation of the inner wall of the pipes. Then, a normal use of the networks has been simulated by passing water at a flow rate of *ca.* 100 L/h. The flow is maintained for 10h each day except during the weekends.

A 50cm^3 water sample is taken on the three settings, every two weeks after a weekend of stagnation. These samples are taken at the valve at the bottom of the setting. Iron, zinc and manganese are measured systematically, after acidification, by atomic absorption. In these conditions, the concentrations measured correspond to the dissolved and precipitated metallic elements.

The water used during all the experiments is a tap water from Paris produced by treatment of the Seine water. A mean composition is given in Table 1 (p.202). This water had stood and was always slightly calcified.

After one year, further samples of galvanised pipes from the same place were taken to observe the evolution of the inner wall of the pipes.

4. RESULTS

4. 1 OBSERVATION OF THE PIPE WALLS

4.1.1 State of the walls immediately after disinfection

Samples have been taken on each setting in a horizontal part immediately before the T fitting. The samples, 5cm long, have been sawn longitudinally to observe the upper and lower parts of the pipes.

On the sample taken on the network treated by permanganate, brown deposits of manganese dioxide are visible, these deposits are present in a greater amount on the lower part than on the upper part (photo 1). They are formed by reduction of the permanganate by metallic zinc (corrosion reaction). Manganese dioxide is also formed by the action of grease used for preparing the threading end of the pipes or to ensure the tightness of the threaded fitting.

The pipe treated with hydrogen peroxide shows important white deposits resulting from a strong attack of the zinc layer by the oxidising reagent. The deposits are composed of zinc corrosion products like hydroxycarbonate.

The pipe treated with sodium hypochlorite shows little or no visible modification after treatment.

4.1.2 State of the walls after one year of operation

Samples of galvanised pipes have been taken, in the same place as the first one, after one year of operation. The threaded bends and brazed bends have been taken too at the same time for examination of the inner wall. The first samples have been sawn longitudinally to observe the upper and lower part of the straight pipes and bends. The inner walls of the three straight elements are covered by a protective layer of zinc corrosion products. Red deposits are visible in the pipe treated with permanganate and resulted from corrosion of parts situated before this horizontal part. It appears the deposit formed in the pipe treated by hydrogen peroxide is thicker than on the other pipes. It covers the zinc walls, of course, but it also covers bare ungalvanised steel parts, such as the extremities of the pipes.

Observing the two types of bends gives rise to the following remarks:

- The threaded bends are hardly corroded, with only some traces of corrosion on the ungalvanised parts, such as the uncovered parts of the threads or the extremities of the fittings. The attacks on the metal are less severe in the case of the pipe treated by hydrogen peroxide than in the others. In this case the zinc corrosion products also cover these ungalvanised parts.

- The brazed bends are very corroded, more so than the threaded bends. In this case the bend taken on the network and treated by permanganate is the most severely attacked.

For all the fittings, the ungalvanised parts are very corroded in the case of the pipe treated with permanganate, less corroded in the case of hypochlorite and least corroded in the case of hydrogen peroxide (photo 2).

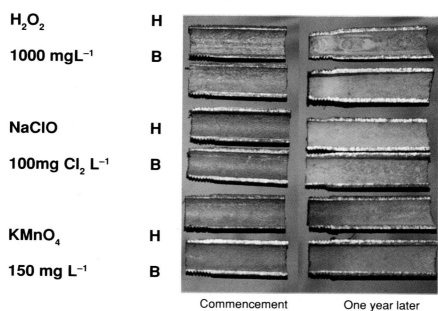

Photo 1 View of the inner walls of straight pipes after disinfection, immediately after and one year after.

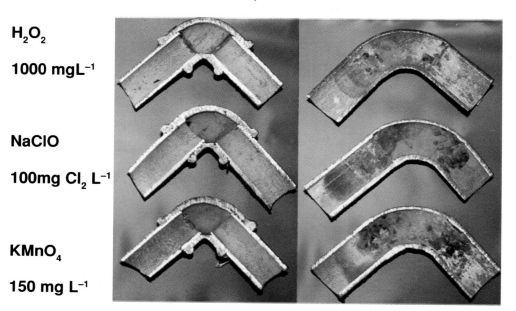

Photo 2 State of the inner wall of the threaded and brazed bends.

Table 1 Characteristics of the water flowing in the settings during the period of the experiment

Parameter	Value	Parameter	Value
Conductivity µS cm^{-1}	450–550	Temp.°C	5–20
Hardness ppm CaCO$_3$	210–250	pH	7.7–7.3
Calcium mg L^{-1}	75–90	Alkalinity ppmCaCO$_3$	170–200
Magnesium mg L^{-1}	3–4	Sulphate mg L^{-1}	30–45
Sodium mg L^{-1}	10–20	Chloride mg L^{-1}	20–30
Potassium mg L^{-1}	2–3	Nitrate mg L^{-1}	15–25

4.1.3 Evolution of iron, manganese and zinc concentrations in the water samples

The results of the analyses performed on each setting are shown in Figs 4–6 and the mean values are given in Table 2. Immediately after going into operation, corrosion of the bare steel parts appeared in the three cases leading to a high concentration of iron in the water samples. But it can also be seen that the concentration of iron in the samples taken on the setting treated by permanganate is considerably higher than in the other cases. In contrast, the iron concentration in the samples taken from the setting treated with hydrogen peroxide is very low and decreased rapidly after the start of operation.

This is related to the presence, in this setting, of the zinc corrosion deposits on the bare steel parts.

In all cases, the level of iron in the sample reaches zero after about nine months, the minimum period of time necessary to get an effective protection of the pipe walls.

For the manganese, a large concentration of this is found in the samples taken from the setting treated with KMnO$_4$ (Fig. 5).

The concentration decreases slowly during the first five months to reach zero. This shows the leaching of manganese from the deposits formed during disinfection. For the other settings the level of manganese stays very low and near the level of the concentration at the inlet of the settings.

For the zinc concentrations (Fig. 6, p.204) the evolution during the experiment is similar from one setting to the other, and shows a slow decrease of zinc

Table 2 Mean values for one year of the concentrations of iron, manganese and zinc in the water samples

Reagent used	Manganese µg L^{-1}	Iron µg L^{-1}	Zinc mg L^{-1}
H$_2$O$_2$	34	641	7.30
HClO	30	1450	6.34
KMnO$_4$	188	2580	11.42

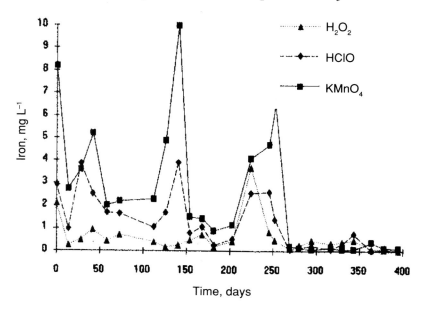

Fig. 4. Evolution with time of the iron concentration of water samples taken after stagnation.

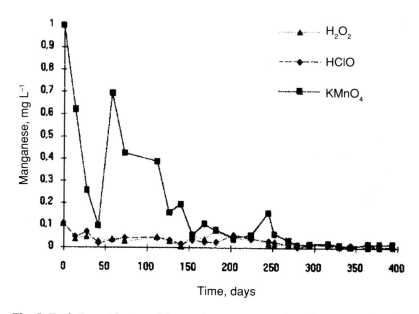

Fig. 5. Evolution with time of the manganese concentration of water samples taken after stagnation.

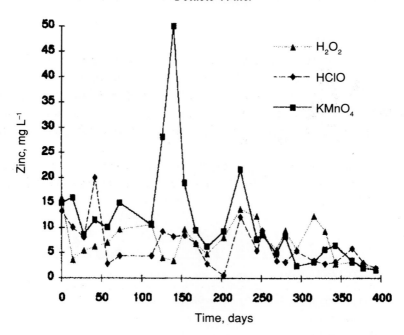

Fig. 6. Evolution with time of the zinc concentration of water samples taken after stagnation.

concentration. But the samples taken from the setting treated by permanganate present a higher level of zinc than the others. The zinc concentration of the water from the setting treated with hydrogen peroxide is somewhat higher than that of the water from the setting treated with hypochlorite.

This results in the activation on the zinc surface by hydrogen peroxide during the disinfection. Part of the zinc ions precipitate on the walls of the pipes and protects them against corrosion.

The experiments described here show that the permanganate remaining in contact with the pipes for only a day lead to the formation of manganese dioxide deposits on the walls of the pipes. This by-product combines with some micro corrosion cells on the zinc surface, increasing the corrosion rate and disturbing the formation of the protective layer.

The treatment with hydrogen peroxide seems to have a strong effect on the zinc surface, destroying the initial passive oxide layer during disinfection. The surface is then activated and corrodes quickly but uniformly. The excess of dissolved zinc precipitates as hydroxycarbonate. After the disinfection treatment, the deposits protect both the metal and galvanised parts, as well as the bare steel parts, and decreases the corrosion rate. The treatment using hypochlorite seems to have no real effect on the behaviour of the pipe.

5. CONCLUSION

This study shows that a very short time treatment can have an important effect on the behaviour of the network over a period of one year or more in certain conditions, particularly if the network has not been well laid. It is clear that the choice of the reagent used for disinfection must be taken into account together with disinfecting efficiency, but it must include the secondary effects on the pipes themselves.

The effect of the disinfection is clearly shown and constitutes no negligible cause for any corrosion of the galvanised networks in the case of use of chlorine or potassium permanganate.

REFERENCES

P. Leroy and J. Teisentz, How Paris feels about the quality of its water. *Water Supply*, 1992, 10, Jonkoping, 47-54-3.

L. Legrand and P. Leroy, *Prevention of Corrosion and Scaling in Water Supply Systems*. Ellis Horwood, London, 1990.

DISCUSSION

Mr Lackington asked whether Dr Leroy ever received complaints about furred-up kettles in Paris, as we do in the UK, where calcium carbonate is deposited. Dr Leroy explained that he did, but with differences. They use two types of water with similar hardness; surface water from the Seine or Marne, and underground reserves. Only the underground water, which is spring water from the South and East, causes scaling. This demonstrates that increasing the hardness with calcium does not necessarily lead to problems with scaling.

Mr Campbell stated that small amounts of organic matter in the water affect scaling tendencies. He had found that organics in surface water promote the deposition of a fine-grained, adherent, eggshell-like scale. Underground water leads to nodular deposits which are less protective. Polarisation curves with eggshell scale display high cathodic polarisation (a virtue of organic matter) whereas nodular scale gives little cathodic polarisation. Dr Leroy commented that organic matter has an inhibiting effect on the nucleation of calcium carbonate.

Dr Chamberlain asked whether Dr Leroy had encountered any problems where peroxide is used to treat water that is subsequently treated by ultraviolet radiation. Dr Leroy replied that peroxide has only been used for experimental purposes. It can only be used in new networks and is disastrous with old networks because it is unstable when in contact with ferrous compounds. For this reason, only chloride or potassium permanganate are used at present.

Mr N Barraclough of Thames Water asked if different concentrations of chlorine, peroxide and permanganate had been used so that the disinfection capability of the 3 solutions were equal; if not, what factors influenced the choice of the individual concentrations? Dr Leroy explained that the normal levels of chloride and permanganate had been used. In practice, these levels correspond to quite an equal disinfection capability. The concentration of hydrogen peroxide used had been too high and could be reduced in practice.

Corrosion of a Welded Steel Pipe for a Potable Water Supply

P. L. BONORA AND L. GHIRARDELLI

Department of Materials Engineering, University of Trento, Italy

1. INTRODUCTION

A case of corrosion in steel pipes for a potable water supply is presented where the synergistic effect of four factors heavily affected the corrosion resistance leading to many premature failures:

1. The potable water was very soft and rich in CO_2 and no action was suitable to reduce its aggressiveness. Table 1 summarises the main features of the water;
2. The coating of the pipe, both external and internal, was very poor, both as regards material (mastics) and application, with defects and poor adhesion in many locations;
3. The average quality of the steel was poor but, worst of all;
4. The longitudinal welding was undertaken with an electrode material which was anodic to the matrix so that a galvanic couple with unfavourable anode to cathode area ratio was obtained.

The backfill of the pipes was a layer of gravels while the soil was mainly made of clay and sand, including may stones of various dimensions, some of which were adjacent to (and sometimes damaging) the asphalt coating of the pipes.

No evidence of stray currents was found along the pipeline.

Some lengths of pipe were cut for analysis both near the corrosion failures and also where the pipe appeared unaffected.

2. EXPERIMENTAL SECTION AND RESULTS FROM VISUAL EXAMINATION OF SAMPLES

Lengths of corroded pipe were sectioned longitudinally and then examined. Figure 1 shows the situation found internally: an iron-rich and powdery deposit corrosion product totally covering and surface.

Figure 2 shows the internal wall after mechanical removal (by light brushing) of the above mentioned products. The bituminous layer, which was applied by the manufacturer, to protect against corrosion was not distributed uniformly in all areas, as evident in Fig. 3.

Table 1 Analysis of the water

pH=7.4	
Conductivity (18°C)	8.4 10^{-5} ohm^{-1} cm^{-1}
Solids contents (180°C)	55mg L^{-1}
CO_2	5.78mg L^{-1}
Ca^{2+}	16.0mg L^{-1}
Mg^{2+}	2.4mg L^{-1}
NO_3^-	2.0mg L^{-1}
Cl^-	1.0mg L^{-1}
SO_4^{2-}	5.0mg L^{-1}
SiO_2	7.4mg L^{-1}
Na^+	2.0mg L^{-1}
K^+	0.7mg L^{-1}

The state of the pipes enabled deduction of the following mechanism; at the points where, for whatever reason (insufficient thickness, discontinuity, microporosity, etc.) water was in contact with the metal two different corrosion phenomena could occur either sub-surface attack of the pipe walls or localised galvanic corrosion of the longitudinal weld.

It was established that while in the first case, the attack had not yet penetrated the second (galvanic attack of the weld) corrosion of the weld metal had already led to perforation of the metal walls (see Figs 2 and 3). In these cases, once the layer of asphaltic insulation was reached thanks to the combined effect of working pressure and capillarity, the water could 'wedge' between the lining and the external metal wall, giving rise to a new under-surface attack (in conditions where perforation had already occurred), as shown in Fig. 3.

The preceding observations have determined the direction of research in three different but complementary directions:

(i) Structural analysis by optical microscopy; (ii) microstructural studies and compositional analysis by Scanning Electron Microscopy and Microprobe (X-ray); and (iii) electrochemical tests.

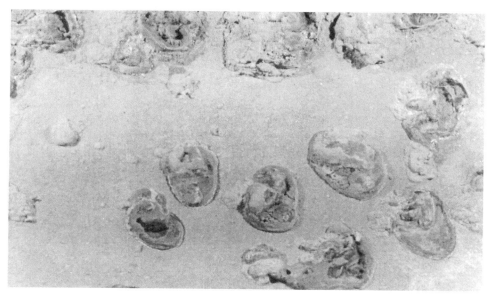

Fig. 1. View of the internal wall of a corroded pipe.

Fig. 2 View of the internal wall of a corroded pipe, after brushing. The perforation due to complete dissolution of the weld is clearly observed.

Fig. 3. Runs in the bituminous coating and lack of fusion.

3. PREPARATION OF THE SAMPLES USED IN THE EXPERIMENTAL PHASES OF THE STUDY

For the structural and microstructural studies samples were obtained from representative cross sections as illustrated in Fig. 4.

Following removal of the insulating material, the samples were ultrasonically cleaned then lightly brushed with a nylon bristle brush to remove powdery corrosion products. They were then mounted and polished to a 1μm surface finish.

At this stage the samples were etched in 2% Nital to reveal the grain boundaries. Those intended for examination by scanning electron microscopy were then gold plated. Longitudinal samples for the electrochemical tests were coated with insulating resins, leaving control surfaces of the weld seam or the walls of the pipe uncovered (approx. 1cm² of surface uncovered in each case).

Figure 5 clearly describes the shape of these samples.

3.1 Structural analysis carried out under optical microscope

The scanning electron micrographs in Figs 6 and 7 show the general sate of the weld in those pipes where it is still present (Fig. 4, case (a)).

In a general visual examination, the weld appeared unaffected in the first case (Fig. 6), whilst in the second it already appeared strongly attacked (Fig. 7). The

Corrosion of a Welded Steel Pipe for a Potable Water Supply 211

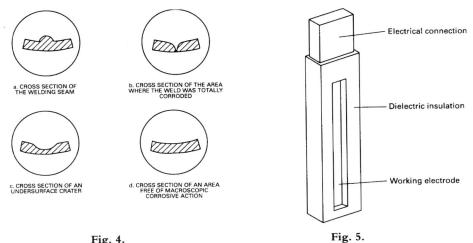

Fig. 4.

Fig. 5.

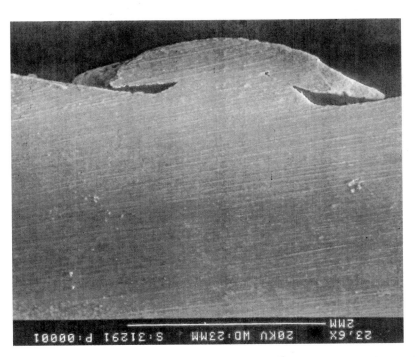

Fig. 6. Micrograph (2mm mark is shown) — example of welding which appeared untouched to the naked eye (note the heavy corrosion under the mushroom head).

micrograph of Fig. 8 shows the state of the weld: the corrosive action gave the deposited material a mushroom appearance. Note the penetration of the attack under the mushroom head.

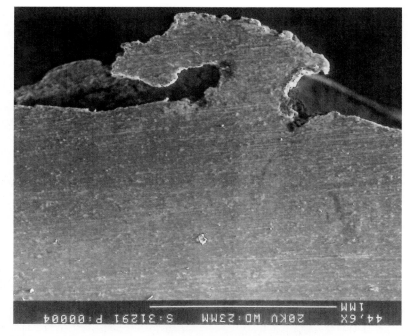

Fig. 7. Micrograph (the 1mm mark is shown). Example of welding which appeared corroded to the naked eye.

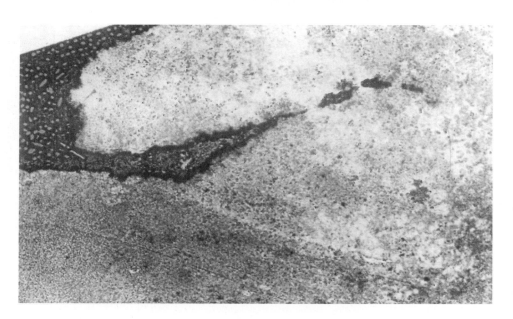

Fig. 8. Micrograph, × 100. Typical sample as in Fig. 4, case (a). Etching in 2% Nital. Detail of corrosion under the mushroom head.

Figures 9 and 10 on different samples, clearly show the structural difference between the deposited metal (Fig. 10) and the steel constituting the pipe walls (Fig. 9).

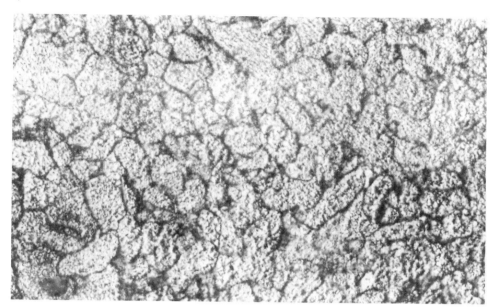

Fig. 9. Micrograph ×1000 — detail of the microstructure of the pipe wall of the sample shown in photo 12.

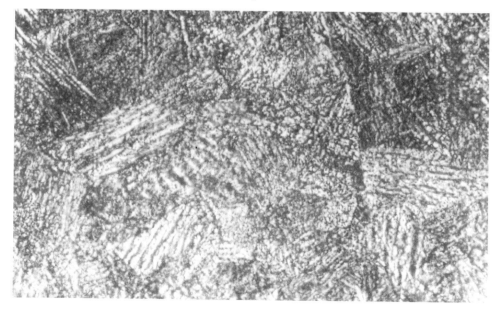

Fig. 10. Micrograph × 1000 — detail of the dendritic microstructure of the weld metal.

Figures 11 and 12 show a sample where the weld was totally removed by the corrosive action (Fig. 4, case (b)).

It is interesting to note that the weld was literally 'dissolved' by the corrosion, leaving the substrate practically intact.

The presence of heterogeneity, constituted by blowholes or small craters (Fig. 12), is particularly evident in the areas next to the weld (completely disappeared), but can also be noted in areas distant from it.

Figure 13(p.216) was taken at the centre of the pipe thickness, in the area below an undersurface crater (Fig. 1), case (c).

The heterogeneity which appears in Fig. 13 is also visible with very similar features in many locations at the centre of the thickness of the steel wall, even on samples of pipe where there were no evident signs of macroscopic corrosive action (Fig. 4 case (d)).

3.2 Structural and compositional analysis carried out under scanning electron microscope and microprobe

In Fig. 14(pp.216 and 217) parts of the longitudinal weld are visible in the area under the mushroom head (see also Fig. 8).

The X-ray analysis, carried out with a microprobe, in the area highlighted, enabled us to obtain a 'map' of the distribution of the manganese and iron in the deposited alloy (See Fig. 15, (colour plate) on p.219): in contrast to a relatively uniform base distribution there were areas of high local values of manganese which at times appear to be real inclusions.

Figure 16 (colour plate, p.219) shows:

(1) The map of manganese distribution;
(2) The map of iron distribution;
(3) The map of ferro–manganese distribution (Fig. 15) at greater magnifications, is identical;
(4) EDXS spectrum representative of the relative local proportions of manganese and iron.

Several samples were lightly attacked with 2% Nital to highlight their structure, before gold plating in preparation for the scanning microscope. The areas of highly localised manganese content were particularly attacked by the reactive acid. In Fig. 17 (p.220) the effect of this preferential attack appears very clearly.

Analyses were carried out on two different samples using X-rays. For each sample (see Fig. 4 case (a)), the manganese and the iron were measured both in the weld and the steel constituting the pipe walls.

The percentage values obtained with the electron microprobe refer theoreti

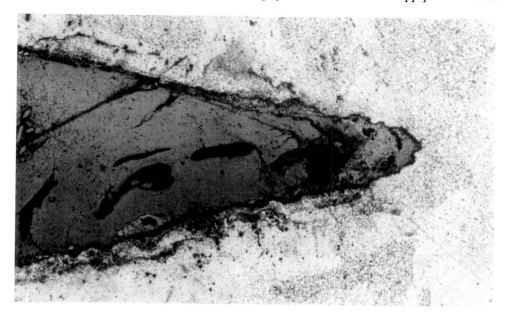

Fig. 11. Micrograph, ×100. Typical sample as in Fig. 4, case (b): detail of the steel microstructure at the edge of the notch where the welding was completely dissolved by corrosive attack.

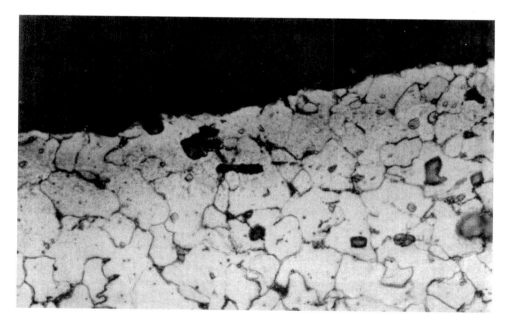

Fig. 12. Defective and heterogeneous structure near the surface (×1000).

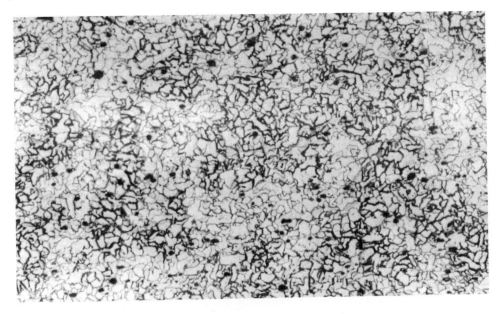

Fig. 13. Defects at the centre of the pipe wall (×200).

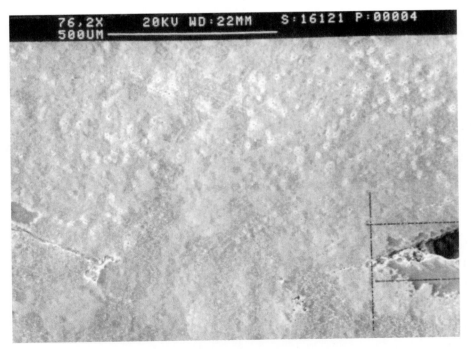

Fig. 14 (a). (Refer to Fig. 4, case (a)) — SEM micrograph (the 500μm mark is shown) of a 2% Nital etched corroded zone under the mushroom head.

Corrosion of a Welded Steel Pipe for a Potable Water Supply

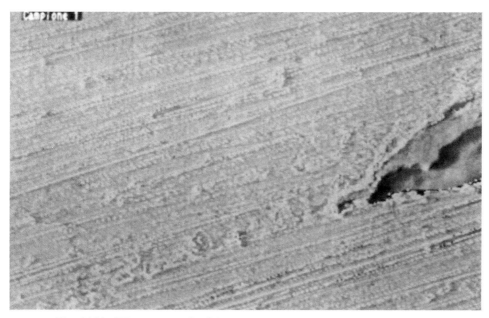

Fig. 14(b). SEM micrograph of the corroded zone under the mushroom head.

Table 2

Sample designation	Steel pipe walls (% Mn)	Inclusions present in weld (% Mn)
Sample 1	0.68	1.25
Sample 2	0.36	1.91

cally to a binary ferro–manganese alloy and are to be considered apart from other components which could be present. Such values are shown in Table 2; the values are in line with what has been described so far.

3.3 Electrochemical tests

Theses tests were carried out to study the difference in electrochemical behaviour existing between the longitudinal weld deposits and the metal constituting the pipe walls (normal steel); for this purpose, the samples described before were used (see Fig. 5):

- Sample 1 = Longitudinal weld,
- Sample 1' = Body of the pipe,
- Sample 2 = Longitudinal weld,
- Sample 2' = Body of the pipe.

Samples 1 and 1' were sectioned from the same length of pipe and samples 2 and 2' from another length; in other words, each pair is representative of the weld and the steel near or adjacent to the weld itself. In both cases the apparent active area is *ca.* 1cm². The following measurements were carried out:

(1) Measurement of polarisation resistance;
(2) Measurement of the direction and intensity of the short circuit current between the pipe body and the weld seam.

In both cases the electrolyte used was 0.3% sodium sulphate in aqueous solution, stirred at room temperature.

This solution has characteristics comparable to drinking water though with a higher electrical conductivity.

3.3.1 Measurement of polarisation resistance

In Table 3, the values obtained for the 4 samples studied are shown.

Table 3

Weld		Body of Pipe	
Sample 1 ohm-cm²	Sample 2 ohm-cm²	Sample 1' ohm-cm²	Sample 2' ohm-cm²
114	128	162	168

As can be seen the polarisation resistance values are higher for the steel of the pipe walls than for the weld.

Measurement of the direction and intensity of the short circuit current.

The samples were placed under short circuit in pairs by means of a nanoammeter at zero internal resistance. The values of the currents measured are reported in Table 4:

Table 4

Pair1/1' µA cm⁻²	Pair 2/2' µA cm⁻²
62	160

The direction of circulation of the electrochemical current, in both cases was from the weld to the pipe body (the weld having a markedly anodic behaviour).

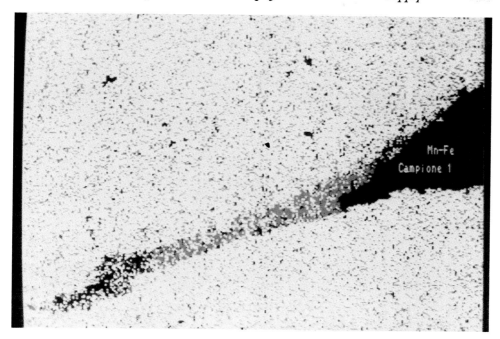

Fig. 15. Manganese and iron distribution map (see Fig. 14).

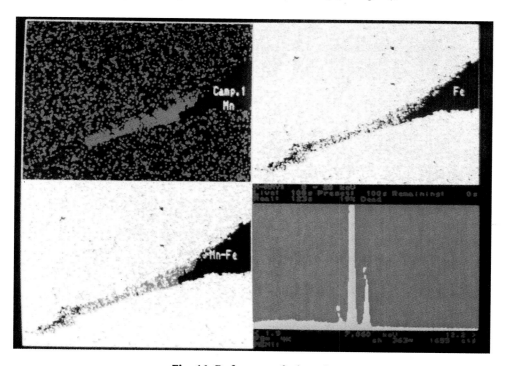

Fig. 16. Refer to text for legend.

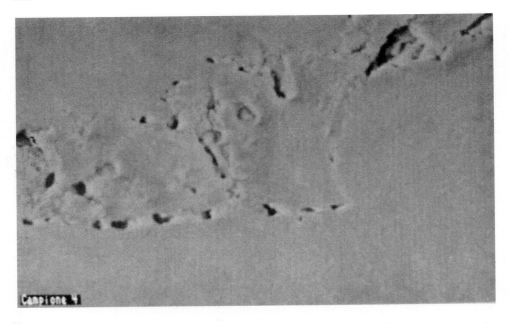

Fig. 17. Preferential corrosion attack by 2% Nital on the Mn rich zones.

4. CONCLUSIONS

The anodic reaction that had begun in a defect in the metal surface (for example, a blowhole), was able to develop and progress by means of the poor barrier qualities of the bituminous coating. This, in fact, did not initially impede the anodic process and actually went on to sustain it because of its own porosity and defective nature.

The craters which developed and which are visible in Figs 2 and 3 represent the final stage in this process of degradation. It is to be noted however that the seriousness of this phenomenon is due in certain areas to the poor quality of the application techniques.

The poor total quality of the protective coating has prematurely highlighted what, in our opinion, is the critical manufacturing defect which has rendered the product unusable. This is the use of a filler material (for the weld seam) with electrochemical characteristics which are decidedly more anodic with respect to the pipe material.

The galvanic cell so constituted has very few possibilities of being polarised (either by the coating or the environment), therefore the corrosion proceeds in a localised manner initially at the seam/pipe body interface, dissolving the whole of the weld as it progresses.

5. ACKNOWLEDGEMENTS

The authors are deeply grateful to Luca Bendetti and Paola Pisetta for their technical support.

DISCUSSION

During discussion of Prof. Bonora's presentation, Dr Dulieu expressed the view that the leaking welds were untypical of the performance of welded steel pipes. Manganese enrichment of welds should not occur and he suggested that there may be something unusual about the welding electrode. He enquired about the chemical compositions of the weld metal and parent plate and also about the specification of the coating since solvent-free epoxies should give excellent protection. Prof. Bonora stated that the manganese was present because of the poor quality of the steel, which was inexpensive and contained many spongy areas full of voids. The coating was also of inferior quality due to inadequate specifications. Such problems are not uncommon in Italy and explain why 30% of mains water in Italy is lost due to leakage. Because of this, Italy is planning to import water from Albania via an aqueduct under the sea at the rate of $4m^3 s^{-1}$.

Dr Kruse stated that weld corrosion is very common in Germany, but only with tubes that have been welded using an electrical resistance method and steel strip with an as-rolled edge. It has been attributed to the enrichment of impurities from the edge of the strip during the welding process. If machined edges are used, the problems disappear and the machining of weld seams was originally suggested in the 1970s.

In response to a question from Mr Pascoe, Prof. Bonora stated that the failed pipe had been in service for only two years. Dr Kruse volunteered that penetration could have occurred within six months if it had been a galvanised steel pipe.

An Unusual Form of Microbially Induced Corrosion in Copper Water Pipes

H. S. CAMPBELL*, A. H. L. CHAMBERLAIN AND P. J. ANGELL

School of Biological Sciences, University of Surrey, Guildford, GU2 5XH, UK

1. INTRODUCTION

The work described in this short paper formed part of a research project, funded by International Copper Association Limited, which will be reported more fully in the appropriate journals. It concerns a novel form of pitting that has been encountered in only five hospital installations worldwide to date. This differs from the more frequently encountered Types 1 and 2 pitting in that it combines certain of the characteristics of these forms of pitting together with evidence of microbial activity. The paper describes investigations conducted at the University of Surrey to establish the role of micro-organisms in the pitting process which has become known as 'Type $1^{1}/_{2}$' pitting.

2. TYPE 1 PITTING

The type of corrosion under discussion should first be distinguished from other, more common types. Figure 1 shows the bore of a copper cold water pipe which failed by 'Type 1' pitting. The mounds of green corrosion product above the individual pits consist mainly of basic copper carbonate, with a general scale on the surface between them of calcium carbonate stained with copper salts. If the corrosion product mound is lifted off, the pit beneath is seen to be roughly hemispherical and to contain loose crystalline cuprous oxide and variable amounts of cuprous chloride. Across the mouth of the pit is a thickened cuprous oxide membrane occupying the position of the original tube surface. The carbonate scale

---

*Department of Materials Science and Engineering.

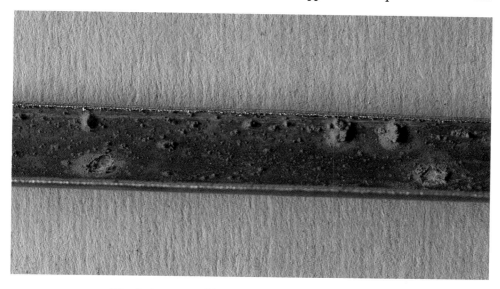

Fig. 1. A copper cold water pipe showing Type 1 pitting.

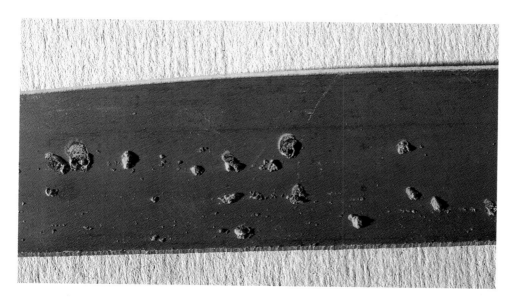

Fig. 2. A copper hot water pipe showing Type 2 pitting.

between the pits spalls off easily showing a thin, shiny cuprous oxide layer beneath. Type 1 pitting occurs in cold water pipes carrying certain hard, organically pure waters and affects only tubes containing carbon films. BS 2871 Part 1 requires copper tubes to be free from such 'deleterious films'.

2. TYPE 2 PITTING

Figure 2 (previous page) shows 'Type 2' pitting. The mounds above the corrosion pits are much smaller and harder than those of Type 1 pitting and consist of basic copper sulphate and oxide. The surface between the pits carries an adherent matt deposit of cupric oxide above cuprous oxide. There is no carbonate scale but a silty deposit may be present. The pits beneath the mounds are of small cross-section and show a branching morphology. They contain hard-packed crystalline cuprous oxide. Type 2 pitting occurs only at temperatures above 60°C in pipes carrying soft waters with a sulphate content exceeding their bicarbonate content. Carbon films are not involved.

3. TYPE 1^1/$_2$ PITTING

Type 1 and Type 2 are the relatively well known types of pitting corrosion in copper water pipes. The type that we have been concerned with is illustrated in Figs 3 and 4. It shows some of the characteristics of Type 1 pitting and some of Type 2. It is often therefore termed 'Type 1^1/$_2$' pitting. The mounds above the pits are relatively large, as in Type 1, but consist principally of basic copper sulphates, as in Type 2. The surface between the pits carries no carbonate scale and is covered by a black deposit of cupric oxide, as in Type 2 but of a more powdery character. There is also usually a deposit of black powdery cupric oxide surrounding the basic sulphate mounds and often partially covering them. The pits are generally similar to Type 1 pits in being approximately hemispherical and containing loose crystalline cuprous oxide beneath a thickened cuprous oxide membrane, but with very small amounts of cuprous chloride.

Type 1^1/$_2$ pitting has been reported only from five hospital installations, one each in Saudi Arabia, Germany and E. Scotland, and two in S.W. England. Neighbouring domestic premises have not been affected. It has occurred principally at 25–45°C and where the service conditions include long periods of stagnation. Both these factors are liable to be present in hospital installations where the water usage is intermittent and there are often long horizontal runs of somewhat oversized cold water pipe running beside hot water services in ducts above false ceilings etc. Although both hot and cold pipes are insulated, the enclosed conditions result in some heating up of the cold water during periods of stagnation.

The waters in which Type 1^1/$_2$ pitting has occurred have total hardness between about 25 and 40mg L^{-1} CaCO$_3$ with alkalinity (carbonate hardness) between 10 and 20. With the exception of the E. Scotland water, they all have chloride contents between 15 and 20mg L^{-1} with sulphate contents between 10 and

30mg L^{-1}. The E. Scotland water has 8mg L^{-1} chloride and 10mg L^{-1} sulphate. For all except E. Scotland water the sulphate content is approximately twice the bicarbonate content and Type 2 pitting has occurred in place of Type 1^1/$_2$ in parts of the hot water systems operating above 60°C.

4. SURFACE FILMS

The presence of a carbon film in tubes showing Type 1 pitting is usually demonstrated by treating a half section of the tube with 25% nitric acid. This dissolves the cuprous oxide and carbonate scale/corrosion product and loosens the carbon which floats as filmy fragments to the surface of the acid. The application of this test to tubes showing Type 1^1/$_2$ pitting does not show carbon films to be present but does result in the detachment of a gelatinous film which sinks to the bottom of the vessel in which the test is conducted. In the first cases of Type 1^1/$_2$ pitting which he experienced — from Saudi Arabia in 1985 — Campbell assumed that the gelatinous films were siliceous. Shortly after, however, he learned that Prof. W. R. Fischer, investigating failures of a similar type in a German hospital, had identified the gelatinous films in them as consisting largely of polysaccharides — indicating possible microbiological activity. Following this, The International Copper Association sponsored research programmes with Profs. Fischer and Paradies, and more recently Dr Wagner, at Märkische Fachhochschule, Iserlohn and with us at University of Surrey to investigate this unusual type of pitting and, in particular, to establish whether it was microbially induced.

The first step at the University of Surrey was to develop staining procedures for characterising biofilms on corroded samples[1] and to apply these to a number of examples of Type 1^1/$_2$ pitting from different sources. The results showed polysaccharides to be present in all cases. The films detached by nitric acid showed a range of morphology from thick and highly mucilaginous to thinner, sheet-like materials with occasional mixed polysaccharide–siliceous films. Extensive SEM studies carried out by Angell confirmed the character of the corrosion products present in tubes showing Type 1^1/$_2$ pitting and revealed the presence of rod-shaped bacteria in association with the corrosion products and exopolymeric material as shown in Fig. 5 (p.227). The SEM examination also revealed (Fig. 6, p.227) that the powdery black cupric oxide adjoining, and often above, the corrosion pit mounds was composed of roughly spherical nodules each with an outer layer of organic material — probably a biofilm of exopolymeric material from bacteria — although they consisted principally of cupric oxide. It is thought that their unusual form probably indicates some microbial influence in their production.

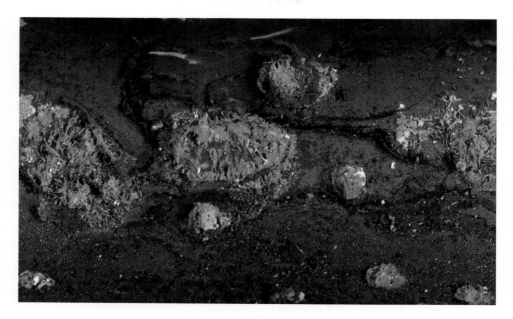

Fig. 3. Cold water pipe showing Type 1¹/2 pitting.

Fig. 4. As Fig. 3 but with one area with intact corrosion products showing 'perforations' in the cuprous oxide pit-closure and adjacent area with corrosion products removed exposing hemispherical pits.

Fig. 5. Bacteria and polysaccharide matrix associated with corrosion products on Type 1 1/2 pitted copper tube.

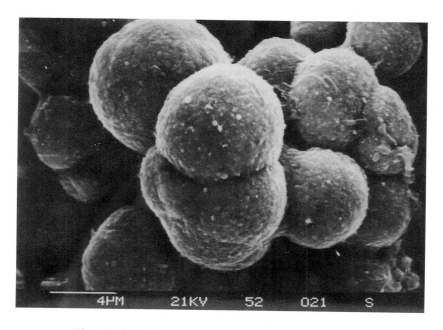

Fig. 6. Spherical granules of the powdery, black cupric oxide.

5. ROLE OF MICROBIAL ACTION IN TYPE 1½ PITTING

In order to demonstrate conclusively that the pitting was being caused by microbial action, it was necessary to approach the problem by applying a slight variation of the 'Postulates' put forward in 1884 by Koch to verify disease-causing organisms. Essentially a single species of organism must be isolated from all animals with a specified disease, grown in pure culture and re-inoculated into a healthy, susceptible animal causing exactly the same disease symptoms, and then finally be re-isolated again.

A 0.25m section of 22mm copper tube was removed from the German installation, refilled with system water and capped before transport to England. The water together with any suspended organisms was removed and plated out onto a range of solidified agar-based media. Swabs of the internal pipe surface were also transferred to broths and subsequently plated out. Eventually the three dominant organisms to appear on the plates were purified, characterised and tentatively classified. All three were species of pseudomonad, including a yellow pigmented strain believed to be *Ps. paucimobilis*, an orange–brown strain also similar to *Ps. paucimobilis* and a species with white pigmentation close to *Ps. solanacearum*.[2] These three organisms were reunited as a defined, mixed culture and used to inoculate a rather novel culture system consisting of a packed-bed of 1cm long 15mm copper tube sections. The systems were established so that two culture vessels ran in parallel but only one was inoculated. The other was maintained sterile to act as a control. Initially, the systems were supplied with a synthetic solution based upon the typical water chemistry described above, but after three months this was replaced by water collected from the S.W. of England where Type 1½ pitting had been identified. Medium was pumped through the systems at 1mL min^{-1} during the day but the feed was stopped for 12h overnight representing the usual hospital scenario. Both dissolved oxygen and pH were continuously monitored and showed that during the stagnant 'overnight' phase the former fell in the inoculated system from 90% to approximately 65%, whereas the sterile system showed no change. The change in pH was minimal in both systems. The temperature followed ambient between 16–24°C.

After approx. 2 years the rings were examined and corrosion observed. There was a very low occurrence of pitting in the 'sterile' system due to the formation of crevices where the ring edges sometimes touched. However, no pitting of the internal surfaces was observed. The inoculated system showed several examples of pit-formation on internal as well as external surfaces. The mounds of corrosion products corresponded in chemistry to those of Type 1½ pitting, although the quantities of the superficial black cupric oxide were very limited and certainly did not form a thick, powdery coating.

Application of the staining procedures detailed above showed the presence of polysaccharide matrix material binding together the microbial biofilm. Much of this extracellular polymeric matrix also gave a positive staining reaction for carboxylated sugars using the alcian yellow/alcian blue technique. This mix of neutral, PAS positive polymers and acidic, alcian positive materials may be very significant if biofilm matrix chemical heterogeneities are as important as is believed.

Subsequently, only the three named, inoculated organisms were recovered from the active system whilst the 'sterile' system produced no growth at all. Thus, a corrosion version of 'Koch's Postulates' has been applied and successfully demonstrated.

A limited survey of copper tube failures from two hospitals in S.W. England showed good correlations between elevated numbers of bacteria, biofilm polysaccharide and Type $1^1/2$ pitting (Table 1). There were a few instances of high bacterial populations with biofilm polysaccharides yet no pitting, but these were invariably in replacement tubes which had been in service for less than 1 year. In no case was Type $1^1/2$ pitting found without the attendant bacterial biofilm. Two cases of Type 2 pitting in hot water pipes acted as an excellent comparison control with no biofilm and very low bacterial numbers.

More recent results suggest that the three organisms isolated and used for the packed bed fermenters are common in cold water systems in hospitals throughout the UK. However, their mere occurrence may not be important, the quantity and distribution of the extracellular polymeric substances is believed to be the key and this will be significantly influenced by the organic and inorganic nutrient levels in the water. Both nitrogen limitation and the presence of copper (II) ions were found to encourage the production of extracellular polymers. Their formation ceased at 45°C and although very limited growth occurred up to 55°C it could not be detected beyond this temperature.

5. CONCLUSION

The survey and laboratory data indicate that Type $1^1/2$ pitting of copper water tube is microbially induced but that water chemistry and mode of operation are also of importance.

Fully detailed reports of the observations outlined above will be given elsewhere in the near future.

Table 1 Correlation of pitting indicators, polysaccharide and biofilm

Sample	Black film	Cu$_2$ beneath mounds	Pitting	Perforation	Film	PAS reaction
Hospital B						
B1	-	-	-	-	-	-
B2	+	+	+	-	+	+/-
B3	-	+/-	-	-	+	+
B4	+	+	+	+	+	+/-
B9	+	+	+	+	+	+
B10	+	+	+	-	+	+
Hospital T						
T9	+	+	+	+	+	+
T10	-	+	-	-	+/-	+
T11	-	+/-	-	-	-	-
T13	-	+	+	-	+	+

REFERENCES

1. A. H. L. Chamberlain, P. J. Angell and H. S. Campbell, *Brit. Corros. J.* 1988, 23, 197.
2. P. J. Angell and A. H. L. Chamberlain, *Int. Biodet.* 1991, 27, 135.

DISCUSSION

Mr A J Graham of ERA Technology referred to two failures that had taken place in Plymouth and in Truro and which were similar to those that had just been described, having occurred in cold/warm water. He asked for a solution to the problem but the authors were unable to oblige. It was suggested, however, that it might help to keep the water moving around to avoid nitrogen depletion. Another idea was to increase the temperature to over 55°C to mitigate the problem. This would, in any case, help to avoid *Legionella* bacteria. It was felt that there was little scope to increase the level of disinfection.

Mr A D Mercer asked whether the phenomenon is restricted to a certain number of hospitals. Mr Campbell confirmed that this is so. Prof. W Fischer mentioned that in Germany similar problems have been seen in hospitals and certain other buildings, including private houses. However, the corrosion rates are much lower in the other buildings than in the hospitals. This means that there is usually no need for home owners to worry about premature failures.

Dr Nuttall asked whether the authors would have expected different results if they had used water from which the organic materials and/or bacteria had been removed by filtration. Dr Chamberlain thought not, if only because they would have been unable to remove entirely all the organic compounds without ultrafiltration or reverse osmosis treatment and removing *all* bacteria simply by filtration is virtually impossible with large volume systems over any realistic time period.

Corrosion in Potable Water Systems — The Situation in Finland

T. KAUNISTO

*VTT (Technical Research Centre of Finland), Metallurgy Laboratory,
P.O. Box 113, SF-02151, Espoo, Finland*

ABSTRACT

The extent and cost of damage caused by leaks in domestic water pipes has been rising during recent years. According to an investigation carried out by insurance companies, the majority of incidents are leaks in piping systems in private houses. The most common cause is corrosion in cold-water pipes. Among the failure cases investigated at the Metallurgy Laboratory of VTT, about half occurred in hot-water copper pipes which developed leaks as a result of internal pitting corrosion. In cold-water piping, hot dip galvanised steel pipes have suffered from local internal corrosion. One factor influencing the risk of corrosion is the water quality in Finland. Fresh waters are typically soft and acid, while the iron and manganese contents are high. Half of the raw water of waterworks is surface water. Surface waters are mostly treated for hygienic reasons, and this treatment reduces the corrosivity of the water. Ground waters are more often of good hygienic quality and, especially in small waterworks, may be distributed without any treatment. There are still many people using ground water from their own wells. According to investigations, 90% of the well waters are corrosive, yet water treatment in private houses is very rare. Although up to 80% of the municipal distribution pipes are now plastics pipes, the water pipes in buildings are mostly made of copper. To diminish the metal dissolution and leakage of water pipes more information should be given to communes, waterworks, constructors and users, and better cooperation between all those concerned is needed.

1. INTRODUCTION

The frequency of damage caused by leaks in domestic water pipes has been rising during recent years. The insurance claims for damage were 78 million FIM in 1983, 274 million FIM in 1987 and today over 300 million FIM. The insurance companies have tried to find out the reasons for leaks by an investigation made at VTT.[1] The data covered a wide variety of incidents involving leaks in water pipes. The majority of damages occurred in private houses which accounted for 57% of the total sum claimed. Leaks in piping systems were responsible for the damage in 76% of the cases and leaks in fittings in 14%. The most common cause was corrosion, the cases being distributed evenly between external and internal corrosion. The sum of claims arising from cases of external corrosion was the highest, followed by internal corrosion as the next most significant cause of leaks. The highest proportion of leaks in piping systems involves cold-water pipes (32%) and the most common type of fitting to develop leaks was the dishwasher (36%), both calculated in terms of repair costs.

VTT is an impartial research centre in Finland and failures of metal components are investigated at the Metallurgy Laboratory of VTT. Corrosion of water pipes, heating pipes and sewage pipes is an area where cooperation between different authorities, design engineers, researchers and users is important. To get more information about the causes of the leaks in water piping systems, the Ministry of Environment, who control the building authorities in Finland, initiated a study concerning the failure cases investigated at the Metallurgy Laboratory.[2] This presentation is based on the report of that study.

2. METALS USED IN POTABLE WATER SYSTEMS

Hot dip galvanised steel pipes were formerly used as cold-water pipes, but nowadays copper is the most common material in water pipes. Plastics pipes have been used for about twenty years in house installations. The cold-water pipes may be of polyethylene, polypropene, polybutene and polyvinyl chloride. Polybutene and the cross-linked polyethylene (PEX) is also accepted for hot-water installations. The municipal distribution pipes are now mainly of plastics.

The quality of plastics pipes, copper pipes and metal fittings is controlled with a type approval system, which is valid in Denmark, Finland, Norway and Sweden. Some products can also get the SFS-mark, given by the Standardisation organisation of Finland (SFS), which guarantees that the products fulfil the requirements of standards. Both quality control systems include an internal quality control and an external one made by an impartial research institute.

Lead pipes have not been used in Finland.

3. WATER QUALITY

Fresh waters in Finland are typically soft and have a low pH-value (< 7), while the iron and manganese contents are high. The surface waters also contain a lot of organics which by disinfection may form organic chlorine components. Although the amount of the aluminium remaining after aluminium sulphate-precipitation also depends on the humus amount, the aluminium content (especially of the distributed surface water) often exceeds the limit value. The acidification of the environment will further lower the pH-value and raise the content of dissolved aluminium.

In Finland half of the raw water is ground water. Ground waters are more often of good hygienic quality and, especially in the small waterworks, may be distributed without any treatment. The ground waters often contain high amounts of carbon dioxide which increase the aggressivity of the water. Only few water sources could be used without water treatment if the corrosion prevention of metal pipes were to be taken into consideration. The waterworks mainly assume responsibility for the hygienic quality of the water and regard the corrosion prevention as a minor matter. If alkalisation became more common in waterworks using ground water the corrosion of the water pipes would diminish.[3]

The instructions concerning the quality of drinking waters contain requirements on the sanitary quality. The limit values mostly follow the EC directive but all of the 60 parameters of the directive have not been included.[3] The technical recommendations contain the pH-value (6.5–8.8) and limit values, for example, for chloride, sulphate and for some metals which could dissolve from metal pipes.

In Finland data relating to waterworks, water treatment methods and water quality has been gathered since 1970. The distributed surface waters have fulfilled the water quality requirements and recommendations, but about 20% of the distributed ground waters have not.[4] The treatment processes are either ineffective or are entirely lacking.

About 20% of the Finnish people are still outside the municipal water distribution and they use ground water from their own wells. 85–90% of the well waters are corrosive, yet water treatment in private houses is very rare.[5]

4. FAILURES IN WATER PIPING SYSTEM

The aim of the investigation carried out at the Metallurgy Laboratory was to identify the main causes of pipe-line damages in buildings based on failure analyses. The causes of the failures were divided into three categories:

- corrosion failures
- mechanical failures, and
- technical failures (manufacture, material and installation failures).

In most of the 190 cases the failures had caused leaks. Some pipes were clogged and the quality of the water had become worse because of the corrosion.

Half of the failures were in hot-water pipes and the amount of failed cold-water pipes was almost as high. Only few cases were in valves, fittings and sewage pipes. In a classification according to the material about 60% were copper pipes and about 30% galvanised steel pipes.

When the material and use are combined the distribution of failures is as follows. Half of the failures occurred in hot-water pipes made of copper. One third of the failures comprised cases with hot dip galvanised steel pipes, and 10% cases with cold-water copper pipes.

About 4% of the failures were caused by mechanical or technical reasons. About 7% consisted of unclear cases. In the rest of the cases the cause of the damage was corrosion.

A considerable amount (40%) of the samples had been in use for five years or a shorter time before the failure. Three quarters of the samples were 10 years old at the most. The mean lifetime was a little less than 6 years. It is clear that the damaged pipes are sent for an investigation only when the lifetime really is shorter than the expected lifetime. Because of the cost, investigations will only be performed if, for example, the customer is considering legal action and needs an impartial statement of the damage.

5. CAUSES OF THE FAILURES

The corrosion damage can be a consequence of defects in pipe material, an incorrect installation, a corrosive environment or/and the operation conditions. The requirements for metal pipes are mainly based on the mechanical properties. Deviations from the requirements in the standards give reasons for claims for compensation, but their influence on corrosion resistance is often difficult to evaluate. The corrosion environment, especially the water inside the pipes, significantly influences the behaviour of the metal pipes.

5.1 COPPER PIPES

Copper is considered to be a very resistant material. After a failure the quality of the pipe is suspected first, yet among the failure cases investigated at the Metallurgy Laboratory a material defect was a possible cause of the failure in one case only.

The copper pipes were leaking (44%) mainly as a result of internal pitting corrosion. Uniform corrosion had occurred in 7% of the cases. Other failures had occurred because of external corrosion (6%), erosion corrosion (14%), stress corrosion cracking (10%) and corrosion fatigue (7%).

Internal uniform corrosion can cause a so-called blue water discoloration. In new installations copper will normally dissolve for some months before a protective corrosion product layer is formed. In soft and acid water the dissolution continues, and the users must remember not to use warm water for cooking or drinking. The recommended limit for the copper content in drinking water is 0.3mg L^{-1}.

The leaks had mostly occurred in hot-water pipes because of pitting corrosion, which mainly is a consequence of an unsuitable water quality for copper. The soft and acid waters often contain aluminium, iron and/or manganese impurities which, after precipitation on the surfaces of the warm water pipes, lead to local corrosion. Even if the water is of good quality at the time of the failure the pipes may 'remember' the earlier worse conditions. If a protective layer has not been formed at the beginning, the corrosion can continue in spite of the better water quality later. The water used in pressure testing can for example be dirty or contain solid particles. If this kind of water is retained for longer periods in the pipes, it can be the cause of later corrosion problems.

Erosion–corrosion is possible when the water flow rate is too high or there is turbulence caused by changes in the flow cross-section of the pipe. As a soft metal copper is susceptible to erosion–corrosion. Most failures often occur in improperly made joints and bends. The real flow rates are sometimes too high because of too effective pumps. The more corrosive the water is, the more significant is the influence of the flow rate.

Alternating mechanical stresses can, together with corrosive water, cause corrosion fatigue. The cyclic load is usually a consequence of differential expansion or pressure impacts from the water system equipment or fittings. Failed pipes were mostly installed in the floor structures, for example in concrete. Copper pipes should be installed so that they can move in accordance with thermal expansion. Corrosion pits are sometimes starting points for fatigue cracks.

Tensile stresses and a specific corrosive agent are the necessary conditions for stress corrosion cracking. The stresses can be internal, originating during manufacture. The corrosive agents are solutions containing ammonia or nitrates, for example washing powders or detergents. Failures had started on the outside surface of the pipes which had been installed under the floor of washing rooms. In these circumstances the copper pipes should be annealed and carry a plastic coating.

Wetness on the pipe surface may result in external general corrosion. The surface may become wet for example because of condensation. Clean condensed water does not corrode copper strongly, but in some cases the moisture may dissolve corrosive substances from the materials lying around the pipes. The external corrosion will be prevented by keeping the pipe surfaces dry. The pipes should be protected by coatings, if the environment is known to be wet.

There have also been failures due to rough handling. Freezing of the water in the pipe has sometimes broken pipes in Finland.

5.2 Hot dip galvanised steel pipes and joints

Local internal corrosion (pitting, deposit and crevice corrosion) caused over half of the failures in the hot dip galvanised cold-water pipes. About 20% of the cases consisted of internal general corrosion and 10% of external corrosion.

Zinc is resistant in hard waters having a pH-value between 7–12.5. Acid and soft waters dissolve zinc and the water becomes bad-smelling and tasting.

Manufacturing defects were the main reason for the local corrosion of the pipes. The quality of the welding seam or coating did not fulfil the requirements of the standards. If the welding seam has local flaws, such as incomplete fusion, or there are flashes inside the pipe the hot dip galvanised coating becomes too thin in these places. The thin zinc coating dissolves easily and the bare steel corrodes in oxygen containing water. A pipe of this quality began to leak in less than 10 years. In some cases chloride or copper dissolved from copper pipes in the same system have caused internal pitting corrosion.

5.3 Brass valves and fittings

The brass valves were suffering from dezincification which means clogging and seeping of water through the porous wall. Although the brass material must be resistant to dezincification according to the type approval requirements, it is not always the case with the products manufactured from type approved alloy.

5.4 Cast iron fittings

Cast iron joints corroded on the outside surface. The damaged joints were in underground installations.

6. PREVENTION OF THE FAILURES

Corrosion of water piping systems causes deterioration of the water quality and leads to many kinds of economic losses. Corrosion prevention would diminish these troubles. Most of the water pipes inside buildings are made of copper and therefore the corrosion prevention of copper pipes is of the greatest importance.

Even if plastics pipes become more common in the future, there will still be a considerable amount of metal pipes in existing buildings.

One way to promote corrosion prevention is to improve the water quality. The improvements of water treatment systems may be difficult to carry out in the hard economic conditions of today, but the problem needs wider discussion. The costs in water treatment would mean savings elsewhere, for example in repairs and also by decreasing the water loss from leaking municipal distribution pipes. Although communes have markedly shifted from metal to plastics pipes there is still a good deal of distribution pipes of metal.

The design engineers have not given enough consideration to the corrosion of metal pipes and in practice people know too little about the water pipe materials and their possible failures to demand effective measures. People using their own wells are in the worst situation. They need proper information and advice in choosing materials for water piping systems and a possible water treatment method.

Even if the water is of good technical quality, failures could still happen due to improper installation or operating conditions. The failures arising from unsatisfactory installations would be prevented by better education and training of the pipe fitters. The house owners and caretakers of the houses need more information about the proper operation conditions.

REFERENCES

1. T. Lounela et al., 1989. Vuotovahingot 1988 [Water damages 1988]. Valtion teknillinen tutkimuskeskus, Tiedotteita — Technical Research Centre of Finland, Research Notes 1045. 61. p + app. 8 p. (in Finnish).
2. T. Kaunisto, 1991. Rakennusten vesi- ja viemariputkistojen vaurioiden selvitys [Investigation into pipe-line damages in buildings]. Espoo, Valtion teknillinen tutkimuskeskus, Tiedotteita Technical Research Centre of Finland, Research Notes 1198. 34 p. + app. 2 p. (in Finnish).
3. Talousveden laatu ja kasittely eraissa EY- ja EFTA-maissa [Water Quality and treatment of the potable water in some EC and EFTA countries]. Suomen Kaupunkiliiton julkaisu — Publications of the Association of Finnish Cities, 1991, 604. 114 p. (in Finnish).
4. Vesilaitokset ja veden laatu vuonna 1987 [Water Quality in Waterworks in 1987]. Publications of the Water and the Environment 39. The National Board of Waters and Environment, Helsinki, 1989, 215 p. (in Finnish).
5. P. Makinen, 1989. Happamoituminen ja hapan pohjavesi haja-asutusalueiden vesihuollon ongelmana [Problems of the water supply of rural areas caused by acidification and acid ground waters]. Publications of the Water and the Environment 38. The National Board of Waters and Environment, Helsinki. 84 p. (in Finnish).

DISCUSSION

Dr L L Russell of Reed International Ltd, in the USA, considered the high incidence of pitting in hot water to be unusual since pitting is normally worse in cold water. He asked what had caused the increase in problems in Finland during the past five years. Ms Kaunisto said that the reason was not known. The water quality had not changed but many private houses had been built recently. Ms K Nielsen of the Engineering Academy of Denmark raised the possibility that the pH of the water had fallen in recent years because of acid rain. Ms Kaunisto agreed that changes like that are possible, but did not think that this was responsible for the increased problems because most of the water is treated to control its pH.

Corrosion and Related Aspects of Materials for Potable Water: A User's Point of View

D. W. LACKINGTON

Severn Trent Water Ltd, Leicester, UK

1. INTRODUCTION

Corrosion of materials has an impact on nearly all activities involving the supply of potable water right through from the source to the customer and on the return leg back to rivers via sewage works.

Whilst the strengths of materials, whether involving steel, iron, copper, plastics, concrete, brick, lead or others, used in the water industry may be adequately assessed at the installation stage, their durability with time is much more difficult to forecast. Furthermore, their impact on water quality and the health of consumers is even more difficult to assess, particularly when faced with ever moving standards imposed by changing legislation.

This chapter addresses problems associated with mains and service pipes, some of which date back to Victorian time but are now the subject of current legislation. Table 1 provides details of the different materials that have been used in the UK for these applications since 1870. To begin with, cast iron mains and lead services were the only materials for which the Distribution Engineer was responsible. Now at least 20 options are available for mains and service pipes. This chapter begs the question of whether today's distribution engineers will pass on a better system to future recipients than the one they inherited from their predecessors and reference is made to experiences of Severn Trent Ltd.

Passing on a better inheritance can prove costly.

Systems deteriorate with time depending upon the materials used, the nature of the environment in which they are laid and the methods of installation. Whilst an inheritance may appear to be performing satisfactorily at the initial time of handover, it can deteriorate during the period of inheritance due to no fault of the recipient who is, therefore, faced with three tasks:

Table 1 Material usage — 1870–2030

Material	1870–1910	1910–1950	1950–1990	1990–2030
Mains				
Cast Iron	+	+	+	* CL
Spun Iron		+	+	* CL
Cast Iron CL		+	+	*
Spun Iron CL		+	+	*
Asbestos Cement		+	+	+
Steel		+	+	+
Ductile Iron			+	CL/R
Ductile Iron CL/ Zinc/PE			+	+
Prestressed Concrete		+	+	*
MDPE			+	+
HPPE				+
GRP			+	+
PVC-U-Imperial			+	+
PVC-U-Metric				+
Services				
Lead	+	+	+	*
Galvanised Iron		+	+	*
Copper		+	+	*
Copper Plastic Coat		+	+	*
Black Poly			+	*
MDPE			+	+

Polymeric C.

Fittings

Mats. Selection

KEY
+: in use, still being laid
*: in use, % - liaision/r&d
CL: cement lining *in situ*
R: Retrocoat

(i) Identify the condition of the system at the time of inheritance, its weaknesses and actions necessary to resolve these.
(ii) Identify the rate at which the existing system will deteriorate during the time of inheritance and, again, look at a programme of rectification.
(iii) Arrange for investigatory effort to ensure materials used on both rectification and new work do not prematurely deteriorate both in the short and long term. This includes measuring the performance of existing materials and using the experience to ensure any previous mistakes are not repeated.

In addition to these three mains tasks, the Distribution Manager is expected to ensure water quality and level of service criteria imposed by current legislation are upheld.

Depressurisation of the system arising from burst mains and services, which in turn are caused by internal and external corrosion, render the distribution system vulnerable to back-siphonage; the consequences of which are obvious and which are the responsibility of water companies under current legislation.

2. MAINS AND SERVICE PIPES

Severn Trent's present distribution system comprises 77% ferrous materials of which the majority are protected with little more than a cosmetic dip of Dr. Angus Smith's coal tar solution. Plastics and asbestos materials account for up to 23% of total length with relatively small amounts of pre-stressed concrete, steel and GRP used for trunk mains and aqueducts (Fig. 1).

Service pipes are mostly of lead and copper in near equal quantities with plastics and galvanised iron making up the rest and accounting for about 20% of the total (Fig. 2).

Up until the 1920s, cast iron accounted for almost all mainlaying, both large and small, with run lead joints and lead service pipes.

Spun iron then took over from cast and, being a stronger material, pipes were of much thinner wall section. This made them more vulnerable to both internal and external corrosion.

Although internal factory linings of cement mortar were available in the 1930s, initially they were not extensively used and in 1950 an eminent team of water engineers reported, via the Ministry of Health,[1] that "Attention has been focused on problems arising from the necessity to replace considerable lengths of water main after only short periods of service".

The report went on to recommend the use of thicker metal and/or the provision of better protection in order to reduce the rate of premature failures.

In spite of these recommendations, the industry continued to lay spun iron with

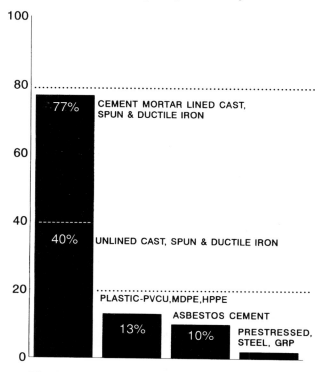

Fig. 1. Severn Trent Water Ltd. Existing mains materials.

the same thickness as before, continued using coal tar protection in all but a handful of cases, added to the 'electrolyte' surrounding the pipe by letting 'unaccounted for' demand grow to around 25% of total demand and then, in the late 1960s, introduced ductile iron which was even thinner than spun iron and, until relatively recently, with no more protection than the usual cosmetic dip of coal tar or bitumen.

Internal corrosion subsequently proved to be a bigger problem than was envisaged in the 1950 report and, in some cases,[2] has been measured as representing twice the problem of external corrosion in spite of the presence of corrosive soils. It is estimated that several thousand tons of iron oxide have been corroded from pipe walls, in areas where water is corrosive, via discontinuities in coal tar and bitumen pipelinings, and deposited within the pipe bore.[2,3] This reveals itself particularly when twin-tub washing machines become filters for the suspended iron oxide.

Redistributed water can react with deposits left by previous water and give rise to customer complaints. If not removed, they represent a major constraint to operational and control activities, and they certainly would if ever the UK water industry shunts water from north to south in a national grid as is advocated by some. The author's experience suggests that only minor changes in water quality

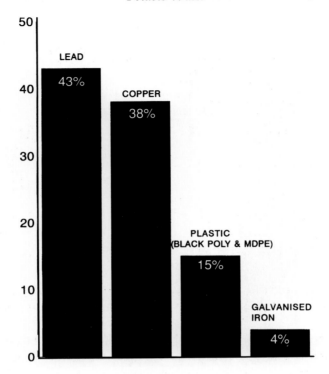

Fig. 2. Severn Trent Water Ltd. Existing service pipe materials.

can react with old deposits and lead to a significant rise in the level of discolouration complaints (Fig. 3).

These same deposits absorb chlorine and, in order to maintain positive residual levels at the extreme boundaries of the system, relatively high dosing rates are necessary at source which generate complaints about taste.

Where the deposits are excessive, they offer constraint to flow and require higher operational pressure to overcome excess frictional loss and maintain satisfactory pressure residuals at the extremities of the system. Higher pressure means higher pumping costs and also gives rise to higher leakage and unaccounted for demand.

Even where waters are not particularly corrosive, the absence of deposits leaves the coal tar linings exposed and in certain cases they react with local water and exceed the limits for polycyclic aromatic hydrocarbons (PAH).

4. THE SYSTEM — 1950–1975

In mitigation, it has to be pointed out that the distribution system doubled in size between 1950 to 1975, a period which saw progressively greater use made of

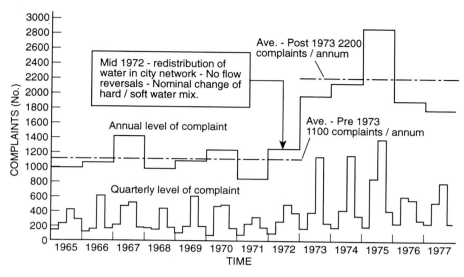

Fig. 3. Discoloured water complaints. Former Leics WD/SOAR Division, 1965–1977.
(Leics WD = Leicester Water Department; SOAR Division = Soar Division of former STW Authority.)

cement mortar, factory lined pipes. The industry also started using plastics, pre-stressed concrete and more asbestos cement pipe. Greater use was made of cathodic protection and thicker bitumen membranes for large mains, particularly steel, and the move towards supplying zinc coatings and PE sleevings to ductile iron pipes during the 1980s represents the most positive action in response to the 1950 report.

The laying of unprotected, thin walled ferrous pipes was not the only source of problem. When PVCu was first introduced in the late 1950s it was oversold and underserviced. The industry was initially more attracted to good deliveries and lower costs offered by manufacturers and paid insufficient attention to quality. Initial claims of good solvent welding properties were soon dismissed due to trench bottom working conditions. Although perfectly achievable in the laboratory, the same performance was not met on site. In addition, premature failure was encountered, particularly with larger mains, due to some early PVCu pipes having inadequate toughness.

Asbestos cement and pre-stressed concrete pipes have not been without problems and the author has personal experience of adverse performance, including biodegradation of natural rubber jointing rings.

5. DISTRIBUTION APPARATUS

The deterioration of apparatus used on distribution systems adds to the industry's problems. Earlier laid sluice valves have similar cosmetic protection to those of mains. Tuberculated deposits are frequently of a magnitude to hinder the closure of shut-off valves, thus making repairs and other maintenance work difficult. Loose jumper hydrants become fixed due to the presence of corrosion, thus impairing their efficiency in preventing potential back siphonage. The effectiveness of air valves is similarly impaired and non return valves can become stuck in the open or closed positions.

6. SERVICE PIPES

Operational repair costs of service pipes have become equally expensive if not greater than those of mains in some areas. This is due in part to the greater intensity of fittings with service pipes having controlling stopcocks and ferrule connections to mains, both of which are vulnerable to corrosion which is exacerbated by the use of dissimilar metals.

As water companies intensify their ACTIVE leakage control efforts, stopcocks which have hitherto lain dormant become the subject of sounding for leak noises and subsequent operation to determine whether escapes of water are on communication pipes (CPs) or supply pipes (SPs). Such operations inevitably lead to damaged spindles, leaking headworks, etc., thus calling for additional maintenance until weak spots are eliminated and the new level of activity becomes the norm.

Although lead is, on average, of greater age than copper, examples exist to suggest that failure incidence of copper is growing at a greater rate than that of lead (Fig. 4). Most of these failures are believed to relate to joints rather than the pipe itself.

The connection ferrules of both tend to become partially blocked with corrosion debris which contribute towards poor pressure problems although the condition of supply pipes is generally the major cause of such problems, as is illustrated by Figs 5 and 6.

Service pipes may be cleaned out using CO_2 or compressed air techniques and Fig. 5 shows how the average flow at the kitchen tap may be enhanced by around 25% following clean out of the communication pipe (CP) in isolation. However, Fig. 6 illustrates how carrying capacity is improved by an average of 100% following additional clean out of the supply pipe.

Most blockage tends to be on lead where the greater presence of common services adds to these problems.

These problems are compounded further by the requirement not to exceed $50\mu g\ L^{-1}$ of lead in water delivered to customers. Whilst treatment of water at source facilitates compliance for most waters, the WHO's recent recommendation involving a $10\mu g\ L^{-1}$ limit would invariably require vast numbers of service

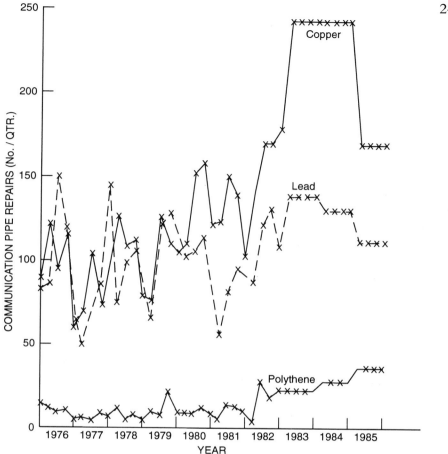

Fig. 4. Communication pipe repairs, former SOAR Division — STW, 1976–1985.

pipes to be replaced. This could create a serious dilemma to water companies whose current responsibility is limited to maintaining only that length of service known as the CP. The remaining length, called the supply pipe (SP) is the responsibility of the customer who appears little interested in its replacement if having to meet the cost. Severn Trent's experience, when undertaking area lead replacement schemes, suggests customer take up of the invitation to replace their SPs is no more than *ca.* 1%.

7. TRUNK MAINS AND AQUEDUCTS

As with stopcocks, if the valves on trunk mains and aqueducts are not the subject of routine maintenance then, after lying dormant for many years, their operation

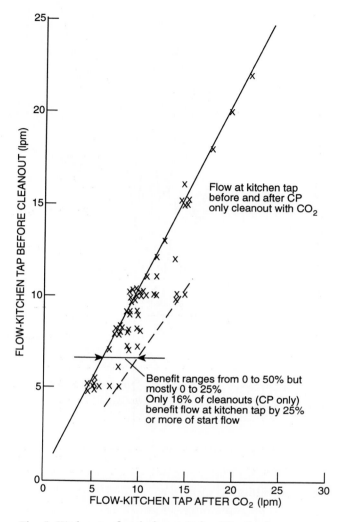

Fig. 5. Kitchen tap flow before and after CP only cleanout.

can sometimes trigger pipe failures due to only minor changes in hydraulic state. Fortunately the burst frequency of large mains is generally much lower than smaller sizes although upstream tuberculated deposits on a trunk main can lead to adverse customer complaint from those connected to the downstream system.

8. LEVELS OF SERVICE AND QUALITY

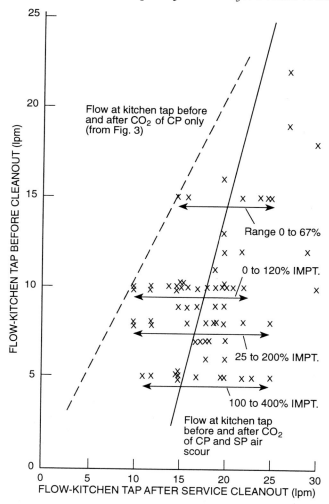

Fig. 6. Kitchen tap flow before and after CP and SP cleanout.

In summary, the distribution system is vulnerable primarily to corrosion of ferrous mains and in particular those that have no internal protection, thus leading to levels of service or water quality problems.

Having inherited 24 000km of unlined iron mains from its predecessors back in 1974, the former Severn Trent Water Authority, for a number of years, maintained a significant mains rehabilitation programme. However, when going private in 1989, new legislation contained in the 1989 Water Act focused the new company's effort even more on customer service criteria governed by:

• Director General's (DG) reference standards
• Internal company standards

- Quality regulations based upon EEC criteria
- The requirements of the 1989 Water Act
- Guaranteed standards — regulations
- Customer warning rules
- Contractor performance and customer care when carrying out remedial work..

Although there are several DG standards against which the company reports, only two relate directly to distribution, namely pressure of mains water (DG2) and interruptions to supply (DG3).

In broad terms, the company is required to report upon the number of properties suffering less than ten metres residual pressure, whilst accompanied by a flow of less than nine litres per minute, at the boundary stoptap if the incident lasts for longer than one hour and is repeated on five days in any twelve months. Alternatively, since the measurement of flow and pressure within service pipes is difficult to establish, an equivalent DG2 criteria is 12m head with no flow in the service pipe.

The DG3 criteria relates to the numbers of properties experiencing loss of supply, without notice, for 12h or more during the financial year under report.

There are exceptions to both DG2 and DG3 reports and incidents are not included in District registers when customers have been prewarned (greater than 24h notice) or when incidents are due to the actions of a third party. In addition, an electricity failure at source represents a further exception for DG3 events.

The company also has a mandatory obligation to comply with quality regulations and prescribed concentrations which relate particularly to distribution include turbidity, iron, PAH, pH, bacteriology, lead and zinc. Failures in relation to iron, turbidity and lead which are associated with distribution systems, are covered by general Section 20 Undertakings by Severn Trent for five years and involve a commitment to prioritise the 17 year programme for eliminating unlined iron mains. Progress on such commitments are audited by the Drinking Water Inspectorate (DWI) under the direction of the Secretary of State.

Severn Trent's in-house levels of service criteria are more onerous than those imposed by the Director General. Four levels of service, namely Excellent (Level 1), Adequate (Level 2), Inadequate (Level 3) and Unacceptable (Level 4) are applied to the following parameters which relate to distribution:

- Pressure of mains water
- Interruptions to supply
- Appearance
- Taste and odour
- Quality of water in distribution.

In respect of pressure, Level 4 is judged to have been reached at 15m head and would trigger renewal or reinforcement action. Similarly a five hour loss of supply occurring five times in any twelve month period is regarded at Level 4 and would generate a reaction.

In order to respond positively and economically in identifying areas for priority action, customer complaints, discolouration problems, repair and maintenance work, customer perception and water quality monitors are all referenced to the

Table 2 Severn Trent Water computer programs in support of rehabilitation priorities

LOSS—D

Pressure logging and analysing system for DG-2 and STW-LOS4 (pressure) violations.

DOJM

Distribution operational Job Management system, although designed for job planning and accountability, contains facilities to:

Identify DG-3 and LOS4 (interruption) events
Identify high spend DMAs
Analyses customer complaints

W-SIR

Measures customer perception of quality, DG-2 and DG-3 levels of service.

ALFI

Automatic Liquid Filtration Instrument - used to identify source of discolouration.

QUIS

Water Quality Information System - records and analyses mandatory and operational water quality samples.

FLOW AND PRESSURE

Supplements W-SIR and LOS-D, measuring flow/pressure at customer tap and local hydrant.

PIPE SAMPLES

Analysis of internal/external pit depths, pipe blockage and forecasts remaining life.

DILIS

Leakage monitor.

company's 2000 District Meter Areas (DMAs). These were originally established in support of the company's ACTIVE leakage control policy. Having determined the length of mains and numbers of properties in each DMA, the comparative intensity of problems can be compared via a suite of computer programs which ensure firm priorities are established. The programs are described more fully elsewhere[3] and are summarised in Table 2.

If water companies are found wanting in their attempts to satisfy water quality criteria, then their licence is at risk and, still worse, individuals are liable to prosecution and imprisonment for up to two years; a sobering thought and a significant incentive for focusing the mind on the important issues when supplying water for human consumption.

9. REMEDIAL WORK

Remedies include scraping and relining using cement mortar, epoxy or thin walled MDPE or complete renewal using either conventional or novel 'trenchless' techniques. on average, the cost of renewal is nearly twice that for relining although the former includes the replacement of communication pipes which accounts for nearly half the difference between the two costs.

Generally the prime triggers for remedial action are as follows:

    Discolouration   —   60%
    Taste and odour —   35%
    DG2 and DG3   —   5%

If all mains were in good condition and of adequate carrying capacity to meet peak demand, then scraping and relining would be the universal solution. This, however, is not the case and mains renewal accounts for between 25 and 30% of the overall rehabilitation programme due either to structural inadequacy, high burst frequency, inadequate carrying capacity or excessive repair costs to mains, services and associated equipment.

Service pipe replacement relies primarily upon conventional open cut or thrust boring techniques although lead liners and displacement techniques are now starting to emerge.

10. SEVERN TRENT WATER REHABILITATION PROGRAMME

During 1991/92 Severn Trent rehabilitated 1867 km of unlined iron water mains and replaced nearly 84 000 service pipes, the majority of which were lead.

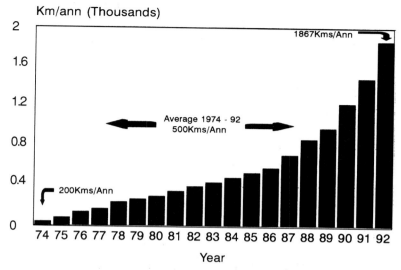

Fig. 7. Severn Trent Water Ltd. Mains rehabilitation 1974–1991/2.

As stated earlier, back in 1974 the then water authority inherited 24 000km of unlined iron mains from its predecessors together with a mains rehabilitation programme of around 200km per annum. During the following 18 years over 9000km were either renewed or relined at an average rate of 500km per annum and of which 4400km have been renovated since privatisation in 1989 (Fig. 7).

The 1991/92 rate was, therefore, nine times greater than that prevailing back in 1974 and nearly four times greater than the average rate since that time. The company is now well in advance of the programmed commitment made when going private to eliminate problematic lengths of unlined iron over 17 years.

11. RENEWAL vs RELINING

The biggest dilemma facing the industry at large is whether to renew or reline and there appears significant merit in pursuing a rehabilitation policy aimed at a much greater proportion of renewal. It would improve the level of customer service, reduce intrinsic leakage characteristics of most systems, reduce maintenance costs, facilitate boundary box metering and, of course, eliminate discoloration problems.

Against this it is argued that burst frequency has only nominally increased during the past 10–15 years and most mains rehabilitation is triggered by discolouration problems that are curable by mains relining which is currently cheaper than renewal. Mains relining also contributes significantly to additional carrying capacity[4] and, therefore, facilitates some pressure reduction and an element of leakage control.

Although the industry has made some progress in reducing mains renewal costs by the development of novel techniques, there is still some way to go before the extra cost is sufficiently small to compare favourably with the subsequent savings in maintenance costs arising from renewing instead of relining.

There is a variety of other influential factors which figure in the renewal vs reline argument and these are illustrated elsewhere.[3] Much will depend on the Government's and customers' reactions to the WHO recommendations on lead, the industry's final response to metering, byelaw enforcement, common service pipes, electrical earthing and leakage policies.

12. EFFORT — CURRENT AND FUTURE

Since the early 1980s, it is arguable that the water industry has responded positively to inherited problems and taken steps to ensure future maintenance burdens and water quality problems are kept to a minimum.

Having experienced failures arising from initial introduction of PVCu and ductile iron, the industry took steps to ensure MDPE was introduced in a controlled manner. A plastic pipes consortium, involving representatives from polymer and pipe manufacturers, WRc, BGAS, water companies and the DTI was established in 1983 to make certain, beyond all reasonable doubt, that the performance and life of the product would match initial claims made by manufacturers.

Fully characterised pipe was buried in readiness for subsequent exhumation in order to test practice against theory and establish some measure of deterioration of blue MDPE with time. In the absence of earlier effort the trials also involved burial of fully characterised PVCu with improved toughness properties to match current day quality. The exercise was later repeated for HPPE prior to introduction in 1990.

The consortium spawned the operational Practices Group which later became the Plastic Pipes Club on which the majority of water companies are represented together with WRc. The aim of both groups has been to pool operational experience and co-ordinate the requirements of research and development. The current MDPE manual relied heavily upon the outcome of information from these two groups.

Although water companies are still learning, particularly in respect of the newer HPPE, there is little doubt that the water industry has benefited from the closer relationships with manufacturers.

The properties of MDPE have facilitated the renewal of mains using novel techniques which have brought economies and reduced significantly the cost gap between renewal and relining. The industry is now able to capitalise on the water carrying hole beneath many pavements in spite of the existing cast iron being beyond a state fit for conventional rehabilitation. Break out and pull through of a liner and MDPE pipe is the preferred option in many cases.

Certainly, the introduction of the new fracture toughness criteria for PVCu has improved its performance and this is reflected in lower burst frequency. As with MDPE, a new manual is available for PVCu.

The same is also true of ductile iron with a similar liaison group (DILG) being established in 1983. Representatives from water companies, WRc and manufacturers have ensured that earlier problems have been faced and remedial measures taken to ensure that the current product, if laid correctly, will have a life in excess of 100 years. The use of factory cement internal linings, zinc coating and PE sleeving is now used by the majority of companies as a matter of routine and where soft water might give rise to pH exceedences, then the answer rests with internal bitumen seal coats. Care is, however, required to ensure that seal coats are cured before release by the manufacturer in order to avoid taste and odour problems.

Where companies have had the misfortune to lay earlier ductile iron without external protection then Retrocat is now available, involving the retrofitting of sacrificial anodes or continuity bonds for impressed current protection; 250mm dia. access holes are made using hydraulic lance and vacuum excavation techniques at a cost equivalent to a fraction of the renewal rate with economies being greatest for the largest of diameters. As part of initial trials, the retrofitting of anodes to mains that would normally have been renewed has seen burst frequency reduced by an average of over 80%.

Water industry liaison groups also exist for other subjects along similar lines to those for PPC and DILG or for pursuing specific R&D activities. Those relating specifically to distribution include:

- Glass Reinforced Plastics — mostly used for large diameter mains.
- Fittings — charged with the task of producing water industry specifications covering couplings, repair collars, flange adaptors, tees, etc., for materials where no satisfactory standard exists.
- Service Pipes — primarily charged with the task of producing a manual on all service laying materials, the laying thereof, maintenance and repair.
- Polymeric Coatings — mainly for the purpose of producing an industry specification (WIS) for thermosetts and thermoplastics together with an accompanying information and guidance note (IGN).
- Material Selection — a manual for mainlaying materials and their selection was produced in 1989 and is currently awaiting update.
- Cement Mortar and Epoxy Lining Materials — charged with the task of evaluating the long term performance and impact on water quality of lining materials.

Most material liaison groups report to the Materials and Standards Group or the Mains and Services Working Group, both of which answer to the WSAs Sewers and Water Mains Committee which in turn is the water industry's focal point for such activities.

In addition to those mentioned above, other groups are pursuing subjects related indirectly or directly to water quality and are all charged with the task of improving the overall service to customers.

13. CONCLUSION

It is, therefore, with confidence that the industry can state it is taking action that

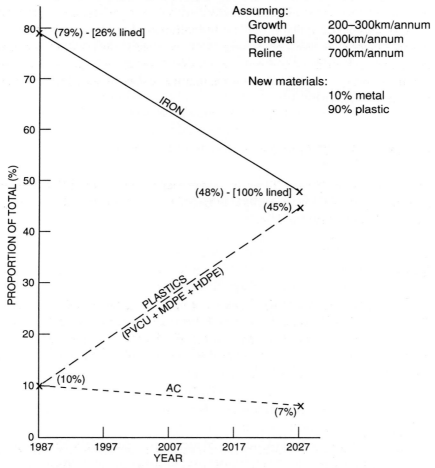

Fig. 8. Proportion of materials in distribution (STW Ltd).

will ensure a better system is passed on to future incumbents. Not only is it facing up to problems inherited from others, it is making greater effort from past mistakes and to study more closely the behaviour and durability of old and new materials before accepting and introducing them into the system. R&D programmes are reflecting the probably changing characteristics of the system over the next 30 to 40 years (Fig. 8).

The cost to the industry in sorting out its inheritance is not cheap and future investment will depend upon legislation, prevailing levels of service, and other demands placed upon water companies for competing capital investment.

The market place will play a part in influencing unit cost and water companies must ensure that a healthy cost benefit emerges from their investments. They cannot 'go overboard' on the philosophy of passing on a better system for future inheritors but they can make sure that investments achieve the desired benefits, and that the product is of good quality and will not prematurely deteriorate.

Whilst appreciating that times are difficult, most water companies wish to join with manufacturers and contractors who see quality of product and customer service as a high priority and non negotiable. This at least goes most of the way to ensuring we get value for money, maintain a quality service and eventually pass on a better system.

14. ACKNOWLEDGEMENTS

The Author thanks Severn Trent for giving permission to publish this paper. The opinions expressed, however, are those of the Author and do not necessarily represent Severn Trent policy.

REFERENCES

1. Ministry of Health, 1950 "Deterioration of Cast Iron and Spun Iron Pipes", HMSO.
2. J. M. Large and D. W. Lackington, 1978 WRc Conference Proceedings, Oxford, England.
3. D. W. Lackington and J. M. Large, *IWES*, 1980, 34, 15.
4. D. W. Lackington, "Leakage Control, Reliability and Quality of Supply", Civil Engineering systems 1991, Volume A, pp. 219-229.

DISCUSSION

Referring to the benefits achieved by cleaning communication pipes, Mr H S Campbell asked how this was done. Mr Lackington explained that this involved removal of the stop tap then blasting through with either compressed air or carbon dioxide gas. However, he stressed that it is no good merely to clean out the communication pipes because most of the problems due to loss of pressure are associated with the common service pipe. Mr Campbell went on to ask whether the cleaning of lead pipes results in an increase of dissolved lead in the water. Mr Lackington stated that this is generally not the case although he finds it hard to believe that this would always be so. Severn Trent Water have 1.2 million customers taking water from 800 000 lead communication pipes, and values of lead in water can range from very low to very high from day to day. There must, therefore, be some risk that the cleaning of lead service pipes would cause approved limits for lead to be exceeded.

Dr D Dulieu, alluding to their advantages over cast iron pipes, asked whether more use would be made of the welded steel pipes in future. Mr Lackington explained that this would depend on the diameter of the pipe. Up to 63mm dia., medium density polyethylene is now used. For diameters from 63 to 200mm, MDPE, PVC or ductile iron are used. For larger pipes, the choice of material depends upon the contractor. Steel can be used satisfactorily if suitably protected.

Study of the Influence of Alkalinity and Calcium on Copper Corrosion

T. HEDBERG AND E. L. JOHANSSON

Chalmers University of Technology, Göteborg, Sweden

1. INTRODUCTION

Water quality criteria in Sweden are based on the analysis of water at the consumer's tap.[1] A range of materials has to be considered when minimising corrosion. While plastics are being increasingly used in Sweden, iron is still used for 60% of water mains, while copper is used in houses and for service pipes because it is considered to have good resistance to corrosion. Leaks due to pitting are the main problem, but interest in general corrosion is greater because the Swedish regulations address drinking water quality, and these regulations can be severely affected by the latter process. Water quality surveys at different places in Sweden have shown that most of the country has soft water, but there is hard water in the south and in a few other places, notably Stockholm. Generally it has been found that soft water areas have red water problems with high copper content in the drinking water.[2-5]

To minimise internal iron corrosion in networks with soft water various corrosion prevention methods are used. The methods are based on addition of sodium hydroxide, carbon dioxide, lime and sodium carbonate in various combinations.

The raw water in Göteborg is a very soft surface water. Consequently, the copper level in this water is low but there are many problems with red water. This water is used for drinking water production at two different water works, Lackarebäck and Alelyckan as shown in Table 1. At the Alelyckan water works, lime and carbon dioxide are added. The resulting increase of alkalinity and calcium level of the water leads to a decrease in the corrosion of iron pipes.

However, the increase of alkalinity and calcium increase the copper concentration at the consumers' taps (Fig. 1).

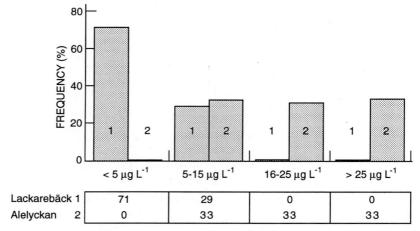

Fig. 1. Copper concentration at the consumers' tap with water from either Lackarebäck or Alelyckan waterworks.

Table 1 Composition of Göteborg raw water, the water from Lackarebäck and Alelyckan water works

	Raw water	Lackarebäck	Alelyckan
Colour mg Pt L^{-1}	20	< 5	< 5
Turbidity FNU	0.83	< 0.05	0.10
pH	7.2	8.1	8.2
Calcium mg L^{-1}	13	15	26
Magnesium mg L^{-1}	1.7	1.6	1.7
Iron µg L^{-1}	63	< 5	16
Alkalinity HCO_3^-, mg L^{-1}	13	15	63
Chloride mg L^{-1}	12	12	12
Sulphate mg L^{-1}	15	26	29

After the household, used water contaminated by copper from the pipes is transported to the sewage water treatment plant and a significant proportion of the copper ends up in the sludge. When the copper content in the sludge exceeds the recommended value, it is not permitted to be used in agriculture. The value today is 600mg kg^{-1} dry solids. Of 348 sewage treatment plants, 33 exceed this value.[6] However, the regulations are to be readjusted to a lower value (200mg kg^{-1} dry solids), which means that just a third will be below the limit.

An investigation from Stockholm showed that 75% of the copper in sludge is the result of copper corrosion in pipes.[7] This is the case even though in Stockholm there are few problems with either iron or copper corrosion. Water quality in Stockholm is shown in Table 2.

Table 2 Composition of drinking water in Stockholm

Colour mg pt L^{-1}	5
Turbidity FNU	0.1
pH	8.5
Calcium mg L^{-1}	38
Alkalinity (HCO_3: mg L^{-1})	57
Iron µg $^{-1}$	10
Chloride mg L^{-1}	12
Sulphate mg L^{-1}	48

The highest level of copper in the sludge (up to 1500mg kg^{-1} dry solids) are found in areas with hard water. Malmö is an example of this (Table 3).

Even though the copper content at the consumers' taps is high, nothing has yet been done to change the water quality. In areas with hard water the consumers also have problems with encrustations as a result of $CaCO_3$ precipitation. To solve these problems softening equipment is installed in household. By this method the calcium in the water exchanges for sodium. Copper corrosion will decreases if the pH increases after softening. However, that method is only possible when the softening treatment is applied in the waterworks.

Table 3 Composition of drinking water in Malmö

Colour mg pt L^{-1}	15
Turbidity FNU	0.2
pH	7.7
Calcium mg L^{-1}	74
Magnesium mg L^{-1}	6
Iron µg L^{-1}	20
Alkalinity (HCO_3; mg L^{-1})	170
Chloride mg L^{-1}	21
Sulphate mg L^{-1}	52

2. EXPERIMENT

The effects of pH, hydrogen carbonate and calcium have been investigated in different studies.[8, 9, 11]

In one investigation the impact of acid rain water was studied. An obvious relation between pH and copper content of water was found as shown in Fig. 2.[8] Low pH values are encountered with water from probate wells in places where there is much acid rain and the bedrock contains little chalk. This leads to high levels of copper in the tapwater and householders are advised to flush their taps to obtain copper-free water and to avoid using hot water in the preparation of food.

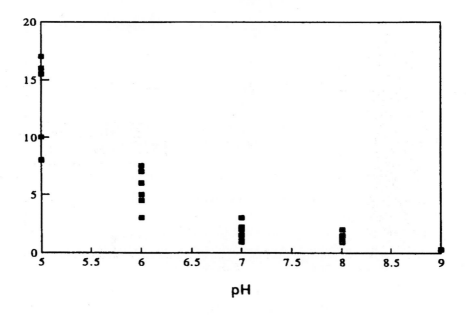

Fig. 2. Copper content vs pH.

Elsewhere, where attempts have been made to overcome red water from iron mains pipes by dosing with lime and carbon dioxide, the copper content of the water has increased. The effect of hydrogen carbonate at 10 and 100mg L^{-1} HCO$_3^-$ at different pH is shown in Fig. 3.[9] The calcium content of the water in theses studies was 5mg L^{-1}. The stagnation time was 20h.

When lime is added to prevent the corrosion of iron the calcium content of the water increases. The effect of increased calcium is shown in Fig. 4.[8,9] The pattern is the same as that found by Pisigan and Singley.[10]

An increase in the calcium content of the water results in an increased copper content and Fig. 5 (p. 264) also provides a very good illustration of the impact of calcium.[11] In one case the water chemistry was adjusted for iron corrosion control by the addition of sodium hydroxide and carbon dioxide and in another case by the addition of lime and carbon dioxide, the latter increased the calcium content from 5 to 15mg L^{-1}. The hydrogen carbonate content was around 60mg L^{-1} in both cases at pH 8.5. The time for stagnation was 20h.

For a more comprehensive view, further investigations have been made using 'hard' water and waters with different calcium contents at different hydrogen carbonate content levels. In Figs 6 and 7 (pp.265, 266) some results are shown. Figure 6 illustrates the copper content for different waters at pH 8.3 with different

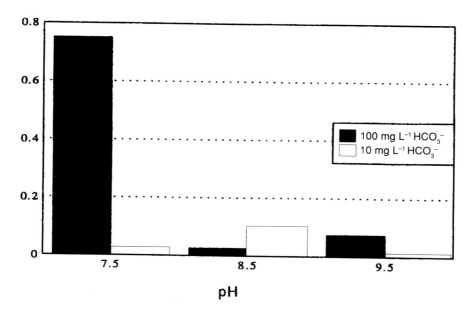

Fig. 3. Copper content at different pH values for two concentrations of HCO.

hydrogen carbonate and calcium contents. The soft water is the Lackarebäck drinking water and the corrosion controlled water has almost the same composition as Alelyckan drinking water (Table 1). The softened water at pH 8.3 has a hydrogen carbonate content of 250mg L^{-1} and the calcium content has been exchanged for sodium. Figure 7 shows three waters with the same hydrogen carbonate content of 250mg L^{-1} at different pH values and calcium contents. Hard water has a calcium content of 80mg L^{-1} and the same pH as softened water, 7.5. The stagnation time was 48h in this study. The general conclusion from Figs 6 and 7 is that pH is the most important factor. It is also very important to adjust pH when the softening is done by $CaCO_3$ precipitation but is more difficult when the softening is done by ion-exchange when only calcium is removed.

3. SUMMARY AND CONCLUSIONS

It is important to prevent copper dissolution from domestic water pipes. The precipitation of various copper compounds can lead to the formation of protective layers at the pipe surface. The copper content of the water increases during stagnation and oxygen near the pipe wall is consumed by the corrosion process,

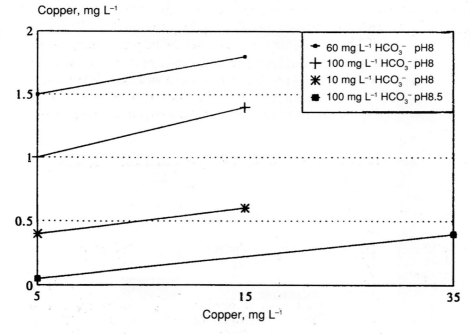

Fig. 4. Copper content at different calcium contents for various HCO_3 concentrations.

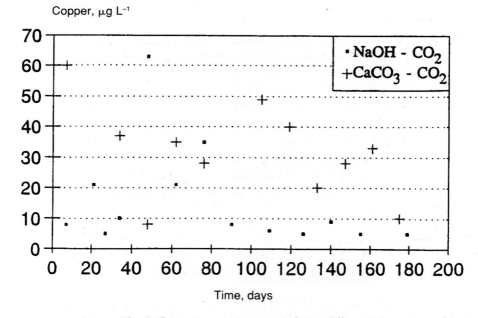

Fig. 5. Copper content in water with two different corrosion control methods.

leading to the precipitation of the copper compounds. When the oxygen concentration is limited near the pipe wall the precipitates may be dissolved according to Hilburn.[12] This exposes the metal surface of the pipe which can be attacked by fresh water after flushing the pipe, as may have occurred in the cases illustrated in Figs 6 and 7. The most protective layers consist of Cu_2O and CuO which form in a hydrogen carbonate free environment when the pH value is between 6.5 and 9 (Table 4). At lower and higher pH the copper ion will be free in the water.

Table 4 Corrosion and protective layer formation situation for different water composition

	pH		
	< 6.5	6.5–9	> 9
Low total carbonic acid content	Corrosion Cu^{2+}	Protection Cu_2O	Protection CuO or Cu(II)
High total carbonic acid	Corrosion Cu^{2+}	Protection $CuCO_3$, $Cu(OH)_2$ Cu_2O or Cu(I)/Cu(II)	Protection CuO Cu(II)

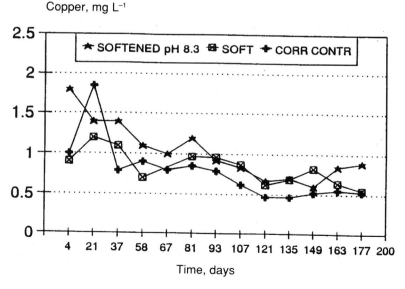

Fig. 6. Copper content for soft water (pH 8.3, HCO_3^- 17mg L^{-1} and Ca 15mg L^{-1}), corrosion controlled water (pH 8.3, HCO_3^- 60mg L^{-1} and Ca 40mg L^{-1}) and softened water 8.3 (pH 8.3, HCO_3^- 250mg L^{-1}) with calcium removed by ion-exchange.

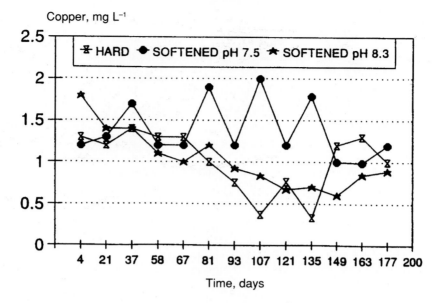

Fig. 7. Copper content for hard water (pH 7.5, HCO_3^- 250mg L^{-1} and Ca 80mg L^{-1}), softened water 7.5 (pH 7.5, HCO_3^- 250mg L^{-1}) with calcium removed by ion-exchange) and softened water 8.3 (pH 8.3, HOC^- 250mg L^{-1} with calcium removed by ion-exchange).

When the water contains hydrogen carbonate $CuCO_3 \cdot Cu(OH)_2$ with less protective properties. The results of such treatments are illustrated in Figs 1, 3, 4, and 5. An addition of calcium may also result in $CaCO_3$ precipitation on the pipe wall. This precipitation is formed as 'nodules' or as 'eggshells' according to Campbell and Turner,[13] which results in high copper dissolution around the $CaCO_3$. Figure 4 shows the effect of calcium addition.

REFERENCES

1. Statens livesmedelsverks kungörelse on dricksvatten SLV FS 1989:30 [Guidelines for Drinking water.] (in Swedish). Uppsala, Sweden.
2. T. Hedberg, Water Quality in the Distribution Network, IWSA, 1982.
3. B. Berghult, A. Elfström Broo, T. Hedberg and E. Lind Johansson, Corrosion of Water Mains with special respect to Iron Pipes. 12th Scand. Corros. Congr. & Eurocorr '92, 1992.
4. S.-E. Kristenson and O. Vergstedt, The redwater problem in Gothenburg. Report from Gothenburg's Water and Swage Works. Göteborg, Sweden, 1992.
5. T. Hedberg, Copper and Iron Corrosion in Drinking Water Water Pipes. National report to the Standing Commitee of Corrosion of IWSA, 1992.

6. Statistiska meddelanden, Utslapp till vatten av eutrofierande ämnen 1990. Statistiska centralbyrån. [Statistic data over the outlet from sewage water works.] (in Swedish) Stockholm, Sweden.
7. Tungmetaller i tappvatten, en förstudie från miljöförvaltningen i Stockholm 1992. [Heavy Metals from Tap Water] (in Swedish). Miljöförvalningen, Stockholm, Sweden.
8. E. Johansson and T. Hedberg, Importance of Water Composition for Prevention of Internal Copper and Iron Corrosion. Dissertation No. 8 Chalmers University of Technology, Dept. of Sanitary Engineering, Göteborg, Sweden, 1989.
9. E. Lind Johansson, Importance of Water Composition for Prevention of Internal Copper and Iron Corrosion. Dissertation No. 8, Chalmers University of Technology, Dept. of Sanitary Engineering. Göteborg, Sweden, 1988.
10. R. A. Pisigan and J. E. Singley, Influence of Buffer Capacity, Chloride, Chlorine Residual and Flow Rate on Corrosion of Mild Steel and Copper, *J. AWWA* 1987, 79, 62–70.
11. E. Johansson, Korrosionskontroll vid Munkfors vattenverk. [Corrosion Prevention at Munkfors Water Works]. (in Swedish). Chalmer University of Technology Dept. of Sanitray Engineering,. Göteborg, Sweden, 1988.
12. R. D. Hilburn, Modeling Copper Corrosion in Water with low Conductivity by using Electrochemical Techniques. *J. AWWA* 1983, 75, 149–154.
13. H. S. Campbell and M. E. D. Turner, The influence of Trace Organics on Scale Formation and Corrosion. BNF Metals Technolgy Centre, MP 596, Oxon, UK, 1980.

DISCUSSION

Prof. Fischer stated that laboratory tests have shown that the blue colouring of water in contact with corroding copper is due to displaced solids, not dissolved ions. He asked whether the authors had determined solids in water and was informed that they had not.

Perforations of Polypropylene Pipes in Potable Water Caused by Cracking — A Case Study

W. R. FISCHER, D. WAGNER, H. SIEDLAREK AND J. A. G. C. SEQUEIRA*

Märkische Fachhochschule, Laboratory of Corrosion Protection, Frauenstuhlweg 31, 58590 Iserlohn, Germany

1. INTRODUCTION

A potable water installation using both copper (SF-Cu, F 37) and polypropylene (aquatherm®) as described in various German standards[1-5] has been in operation now for about five years in an official building in Germany. These materials are used in both the cold and hot water section of the installation. The soft, unbuffered potable water in this water distribution area is supplied from a reservoir (pH 6–6.5) and is alkalized to pH 9–9.5 with slaked lime. Disinfection is achieved by the addition of a mixture of chlorine and chlorine dioxide at the waterworks supplying the building. A quantitative analysis of the water is in Table 1. In this building the uptake of copper into the potable water has been observed in a sporadic, localised and non-regular fashion.[6]

Damage to the hot water section of the polypropylene pipes which contain graphite as a filling agent, has led to 71 failures to date, the first after four years of operation.[7]

This is an unexpected outcome because thermoplastics, including polypropylene, possess a number of hypothetical advantages for corrosion control:

(i) chemical and solvent resistance;
(ii) weight reduction;
(ii) design flexibility;
(iv) specific performance properties; and
(v) costs.

---

* Estrada de Benfica 457-4 Esq, 1500 Locality Lisbon, Portugal.

Table 1 Analysis of a representative water sample taken from the hospital

Conductivity	µS cm^{-1}	123
pH		9.13
Ca^{2+}	mg L^{-1}	12.9
Mg^{2+}	mg L^{-1}	2.71
Total water hardness	°dH	2.4
$Fe^{2+/3+}$	mg L^{-1}	0.08
Mn^{2+}	mg L^{-1}	0.02
NH_4^+	mg L^{-1}	0.03
Na^+	mg L^{-1}	5.96
K^+	mg L^{-1}	0.95
$Cu^{+/2+}$	mg L^{-1}	0.14
NO_2^-	mg L^{-1}	0.01
NO_3^-	mg L^{-1}	7.8
Cl^-	mg L^{-1}	11
SO_4^{2-}	mg L^{-1}	16
PO_4^{3-}	mg L^{-1}	0.13
O_2	mg L^{-1}	6.3
Oxidisability as O_2	mg L^{-1}	0.9
as $KMnO_4$ consumption	mg L^{-1}	3.5

In fact, polyolefins being non-conductors, are not electrochemically and galvanically attacked and are resistant to materials such as acids, bases or salts and even to constituents found in natural soils. Due to these characteristics polypropylene has proved to be very efficient. This is demonstrated by its use for the transport of chemical wastage and hot chemicals.[8-12] Based on this knowledge the occurrence of this damage in a hot potable water installation is very surprising.

This chapter will deal with the failure analyses of the damaged polypropylene pipes taken from the hot water system in this hospital. The results will be discussed in terms of corrosion and ageing effects.

2. EXPERIMENTAL

Samples of polypropylene pipe have been cut between the 3 o'clock and 9 o'clock position, and inspected by optical microscopy. For analysis of not only the surface but also the interior of the plastic material, the samples were cut into small parts. These parts were mounted in resin and polished mechanically with 80 to 4000 grit emery paper. Final polishing was performed with 1µm diamond paste. The results were recorded photographically at different magnifications using optical microscopy.

3. RESULTS

An general view of the characteristics of the failure responsible for a perforation of the pipe is given in Fig. 1 showing a sectioned polypropylene pipe with a crack in longitudinal direction. Figure 2 shows (at ×16 magnification) that the whole inner pipe surface is brittle and full of cracks. The left side appearance of a main crack is depicted in Fig. 3 (at ×16 magnification) while Fig. 4 (p.272) shows the middle of the same crack (at ×32 magnification). The cavity in the middle seems to be the origin of the crack and also shows side-cracks. Furthermore, it can be seen in Fig. 5 (p.272) at ×16 magnification that cavities are formed within the material which also displays pseudo-crystalline embrittlement of the surface. The cross-section in Fig. 6 (p.273) shows clearly that these brittle failures are distributed over the whole pipe surface and not an isolated occurrence. It can also be deduced that these brittle failures are the cause of the perforation of the pipe. Figure 7 (p.273) shows the cross-section of a perforated pipe segment.

The rupture edges seem to be turned to the outside like lips. The foreground of Fig. 7 show that brittle failure initiates in the middle of the cavity while the original inner pipe surface can be seen in the background of this figure.

Fig. 1. General view of a polypropylene pipe (50 ×8.4 mm) showing a failure in the longitudinal direction.

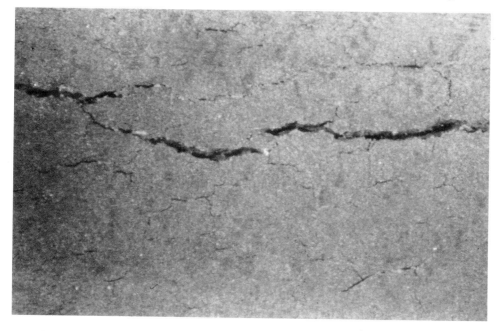

Fig. 2. Surface of the above pipe at ×16 magnification showing a brittle cracked surface.

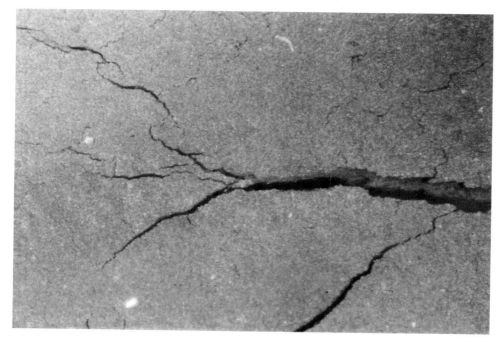

Fig. 3. End of a crack showing ramifications at ×16 magnification.

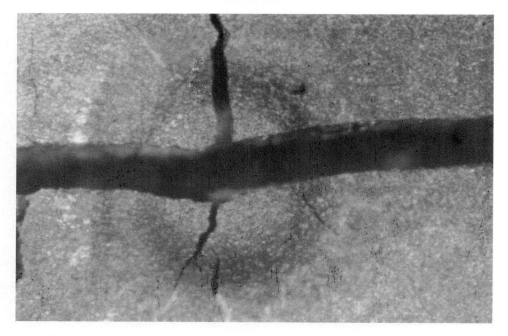

Fig. 4. Middle of the same crack of Fig. 3 at ×32 magnification showing a cavity as the origin of the crack including side-cracks.

Fig. 5. Internal view of a crack at ×16 magnification showing cavities within the material and a pseudo-crystalline embrittlement of the surface.

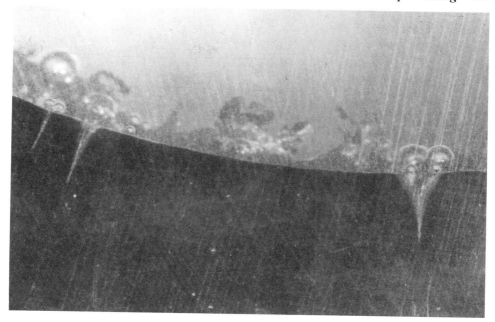

Fig. 6. Cross-section of a polypropylene pipe (thickness 8.4 mm) showing brittle failures.

Fig. 7. Cross-section of a pipe showing a brittle failure leading to the perforation (thickness 8.4 mm).

4. DISCUSSION

The results will be discussed in terms of (i) corrosion, and (ii) ageing of the material in accordance with the following definitions.

The term corrosion, according to DIN 50 900,[13] is defined as a reaction between material and the medium driven by a chemical non-equilibrium condition established between the material and the medium. A special feature is the exchange of a chemical substance across the material/medium phase boundary. The reactions involved may be chemical or electrochemical reactions (stoichiometric formula) or an exchange of at least one chemical substance. These reactions cause measurable changes of the material, i.e. manifestations of corrosion. In the case of metals electrochemical reactions dominate the corrosion behaviour under ambient conditions. Examples of corrosion reactions in polymeric materials are:

(a) The absorption of chemical substances from the medium. This reaction may cause swelling or softening, and, in combination with mechanical stresses stress corrosion cracking.
(b) Leaching of components out of the polymeric material. This reaction may give rise to corrosion cracking.
(c) Oxidation that causes a decrease in molecular weight and, consequently embrittlement. The oxidising agent is then a part of the surrounding medium.

The term ageing is defined as a reaction within the material driven by a chemical non-equilibrium condition established within the material. These reactions might either be diffusion or a phase reaction. A special feature is the fact that there is no exchange of chemical substances across the material/medium phase boundary responsible for the ageing reaction. These reactions cause measurable changes of the material, i.e. manifestations of ageing. These manifestations cause a decrease of the performance of the entity under consideration. Examples of ageing reactions in polymeric materials are (i) degradation of the polymeric molecules; and (ii) delayed crystallisation of partially crystalline thermoplastic material.

Based on these terms 'corrosion' and 'ageing', the results can be explained with the flow diagram in Fig. 8.

The arrows show the possible steps that can lead to the perforation of a polypropylene pipe in chronological order. The dotted arrows show interdependence between the steps that support or facilitate the ageing or corrosion process.

Ageing of polypropylene can begin during manufacture because cooling from the outside leads to the development of different crystallisation grades within the material. The possibility of a non-homogeneous distribution of the stabiliser and

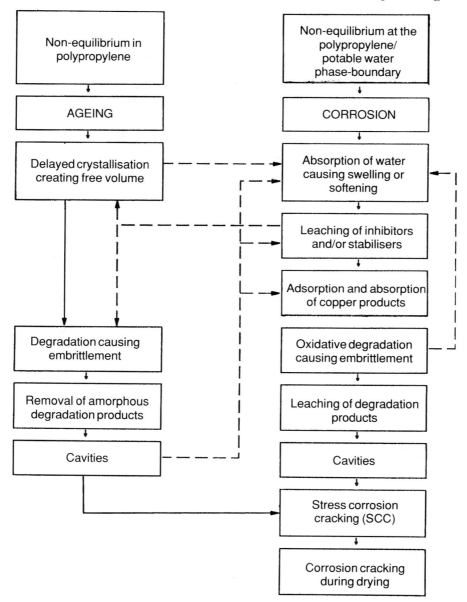

Fig. 8. Diagram showing the possible steps, including interdependencies in chronological order, that can lead to the perforation of a polypropylene pipe.

established within the material before commissioning. Further crystallisation of the polypropylene after the commissioning of hot water systems is facilitated by the high temperature of the water. This ageing process enhances the formation of polypropylene after the commissioning of hot water systems is facilitated by the high temperature of the water. This ageing process enhances the formation of different crystallisation grades, i.e. amorphous and crystalline, and enables swelling of the amorphous phase via the creation of free volume. This can lead to degradation of the material causing embrittlement. The possible removal of amorphous degradation products of graphite used as filling material by washing out processes leads to the formation of cavities.

Cavities can also be formed via corrosion of the polypropylene. During contact with water, a non-equilibrium condition is established at the polypropylene/ potable water phase boundary. This is a prerequisite of corrosion. The absorption of water into the material may cause swelling or softening. Water can also serve as an extraction agent for the stabilisers and inhibitors, leading to the depletion of these substances by leaching. This causes a concentration gradient that lowers the activation energy for degradation of the polymeric material. Copper products present at stochastically high concentrations in this water distribution system [6] can be adsorbed or absorbed by the polypropylene and act as oxidising agents, leading to oxidative degradation and causing embrittlement. The degradation products can be leached out by water and as a consequence, cavities will be formed.

The polypropylene is exposed to both residual stresses and tensile stresses caused by the self-weight of the pipe filled with potable water. Additionally, the processes described above support further embrittlement of the material and the formation of cavities. These factors lead to the formation of cracks by stress corrosion cracking that look like typical brittle failures. These brittle failures can be small in depth or lead to the perforation of the material. After removal of the material from the potable water installation further corrosion cracks may occur due to the drying processes.

Similar effects to those shown in Fig. 5 are described in the literature [14] and are attributed to the influence of a chemical on the polypropylene surface leading to embrittlement and the formation of numerous parallel cracks. This is one reason to believe that over the main perforation and ever throughout the pipe there is a chemical reaction involving the dissolution of the component and crystallisation of polypropylene in parallel.

5. CONCLUSION

If corrosion reactions of metals are considered, chemical reactions and exchange reactions do not usually cause damage under ambient conditions. This rule of practice is not applicable to plastic materials.

Chemical reactions and the exchange of chemical species across the phase boundary determine the corrosion behaviour of polymeric nonconducting materials in aqueous environments. In the presence of stress such processes can lead to stress corrosion cracking. This has been responsible for the failure of polypropylene pipes in a potable water installation also containing copper pipes. In this particular case, ageing and oxidative degradation processes also contributed to the failures.

REFERENCES

1. DIN 1786, part 1; 'Installation Tubes of Copper.'
2. DVGW (Deutsche Vereinigung des Gas-und Wasserfachs)—Working Sheet GW2; 'Capillary Soldering of Copper Tubes for Gas and Water Installations'.
3. DVGW-Working Sheet GW 392;'Copper Tubes for Capillary Soldering in Gas and Water Installations; Rules and Test Determinations'.
4. DIN 1988; 'Drinking Water Distribution Systems on Building Sites; Technical Rules for Building and Running'.
5. DIN 8078; 'Tubes Consisting of Polypropylene, Rules and Test Determinations'.
6. D. Wagner, W. R. Fischer and H. H. Paradies; *Werkstoffe und Korros.* 1992, **42**, 496–502.
7. H. Armbrecht, personal communication.
8. R. M. Burns and W. W. Bradley, 'Protective Coating for Metals', American Chemical Society Monograph Series, Reinhold Publishing Corporation, New York, Amsterdam, London, 1967.
9. *Managing Corrosion with Plastics*, Volume V, NACE Publications, Houston, TX, 1983.
10. J. T. N. Atkinson and H. Van Droffelaar, *Corrosion and its Control*, NACE, 1982.
11. M. G. Fontana and N. D. Greene, *Corrosion Engineering*, McGraw-Hill, New York, 1978.
12. NACE Publication, *Corrosion Basics—An Introduction*, NACE, Houston, TX, 1984.
13. DIN 50 900, part 1; 'Corrosion of Metal, Terms, General Terms', 1988.
14. L. Engel, H. Klingele, G. Ehrenstein and H,. Schaper, 'Rasterelektronenmikroskopische Untersuchungen von Kunststoffschäden', Carl Hauser Verlag München, Wien, 1978.

DISCUSSION

Prof. Bonora disagreed with the view expressed by Prof. Fischer that corrosion can be a chemical rather than an electrochemical process, as exemplified by hydrogen induced cracking of steels. Prof. Fischer maintained that the term 'corrosion' is commonly used in the UK and the USA to describe processes such as liquid metal embrittlement as well as processes involving exchange reactions between gaseous hydrogen and absorbed hydrogen.

Prof. Fischer was asked about the exact temperature at which cracking had caused the perforation of polypropylene pipes in potable water. He replied that the temperature of the hot water system in the official building where the incident had occurred was well maintained at above 55°C but below 65°C.

Index

The inclusion of a page reference in this Index does not imply that a full discussion of that topic will be found. In many cases the item is mentioned merely to set it in the context of a particular technical area.

In other cases, subjects which are included in most of the papers and which are an essential part of the purpose of the book are not indexed in detail; such subjects include, for example, water composition, alkalinity, corrosion prevention, etc.

Acid rain
    possible effects of, on corrosion 239, 261
Ageing
    of polymeric materials 274
Aluminium
    in water supplies, in Finland 235
Aluminium oxide hydrates
    colloidal solutions of, as corrosion inhibitor 62
Aluminium sulphate
    in water treatment 160
Ammonia
    occurrence from nitrate reduction 169
Angus Smith
    coal tar solution, reference to use of 242
Aqueducts
    materials for 242, 247
ARC/INFO
    geographic information system 186
Asbestos cement pipes
    deterioration of 166
    reference to use of 245
Assimilable organic carbon (AoC)
    levels in water 73
Austenitic stainless steel, *see* Stainless steels
Avoidance
    of corrosion problems 114
    of deterioration of cement based materials 166

Badenoch report
    recommended temperatures given in 74
Bimetallic effects 117
Biofilms 10, 73, 83, 115
Brass fittings
    dezincification of 237
    pick up of lead from 91
Bronze
    pick up of lead from 93
British Standards
    reference to BS 6920 (tests for materials in contact with water) 9
Brittle fracture
    of polypropylene 272

Cleaning
    of pipe systems 246
Concrete
    chemistry and degradtion of 48
    constitution of 46
    deterioration of 166
    reinforcement of, by fibres 48, 166
    reinforcement of, by steel 48
    uses of, in potable water 45
Copper
    anions, effect of, 122
    chlorine in water, effect of 69
    content of, in sewage sludge 262
    in water, values 5

corrosion of
    incidence in Finland 235
    in Scotland 141
    in Severn Trent 242
    in Sweden 259
    prevention of 237
corrosion–erosion of 68
deposit attack of 69
dissolution of 67
flux, attack of 70, 82
history of use, in potable water 65
hot softwater, pitting by 68
Legionella, effects of 78
manganese in water
    effect of 69
microbially induced corrosion of 70, 222
operating conditions, effect of, 122
pH, effect of 122
pitting of 122
    by Type 1 mechanism 70, 222
    by Type 1½ mechanism 224
    by Type 2 mechanism 68, 224
    by Type 3 mechanism 69
CPD (Construction Products Directive) 11

Deposits
    absorption of chlorine by 244
    effect of, in Type 2 pitting 224
    in water supply systems 243
    on corrosion 244
Design
    and avoidance of corrosion 115
Disinfection
    in household systems 195
    on site 117
    at waterworks 268
Distribution apparatus
    deterioration of 245
Dolomitic limestone filter
    use of 68
Ductile iron
    liaison group 255
    use of 20, 243
Duplex stainless steels (*see also* stainless steels)
    use in potable waters 105

Electrocxhemical studies
    of corrosion
        of copper 123, 142
        of steel pipe 217
European Commission
    DG XI 16
    Directive Relating to the Quality of Water Intended for Human Consumption (80/778/EC) 67
European Regulations for Drinking Water 61
Experimental studies (*see also* Electrochemical studies)
    of corrosion 208, 261, 269

Faraday cage
    in corrosion tests 123
Ferritic stainless steels (*see also* Stainless Steels)
    resistance to chloride induced stress corrosion cracking 105
Finland
    corrosion in domestic pipes in 232
Flux
    corrosion caused by 70, 101, 118
    phosphate based 101

Gallic acid, *see* Organic matter
Galvanic effects, *see* Bimetallic effects
Galvanised steel (*see also* Zinc)
    contamination of drinking water by 60
    corrosion protection of, in potable waters 62
    damage, types of 58, 237
    disinfectants, effects of 195
    effect of, on water quality 169
    perforation of 60
    temperature of water, effect on 60, 63
    tuberculation of 58
GRP (glass reinforced plastics)
    hydrolytic degradation of 32
    in potable water supplies, use of 29, 36
    testing of 36
Germany
    corrosion problems in
        hospital installations in 224
        official buildings in 268

Gun metal
  leaded, pick up of lead from 93

Halogenated biocides
  effect of, on stainless steels 106
Humic materials (*see also* organic matter) 75, 160, 244
Hydrogen peroxide
  as disinfectant, effect on corrosion 195
Hypochlorite
  effect of
    on corrosion of copper 144, 150, 200
    on corrosion of galvanised steel 105
    on stainless steel 105

Inhibitors
  use of 62, 164

Koch's postulates
  corrosion version of 229

Lead
  carbonate, role in dissolution of lead 85
  content of
    influence of stagnation time on 168
    in water, values 5, 16, 32
  WHO guidelines on 192
  contamination from
    brass fittings 91
    bronze fittings 93
    pipes 85
    stabilised PVC 16, 41, 44
  failures
    incidence of, in Severn Trent 246
  flaking (*see* Lead, particulate)
  particulate 91
  phosphoric acid treatment of water to
    prevent dissolution of 90
  pipes, repairs to, in Severn Trent 90
  sources of, in potable water 85, 168
  toxicity of 65
Leakage control
  practices in 148
Legionella pneumophila
  toxicity of copper towards 66, 77

Loch water
  electrochemical tests in 150

Materials in use
  for potable water supples in UK 241, 255
  tabulation of 255
MIC (microbially induced corrosion)
  of copper 70, 228
  of stainless steel 102
Microbiological assessment 15
Micropollutants
  regulations concerning 5
Migration
  assessment 14
  from metallic products 15
Manganese
  in steel, in relation to corrosion 214
  in water, effect on corrosion of copper by 69

Nickel
  leaching of, from stainless steel 106
Nitrate (in waters)
  cathodic reduction of, on corroding zinc 169
Nitrite
  cathodic reduction of, on corroding zinc 169

Operational factors
  effect of, on corrosion 119
Organic matter (*see also* humic acid, polysaccharides)
  effect of, on corrosion of copper 144, 154, 158, 225
Oracle data base 173
Organoleptic assessment 12
Oxygen (dissolved)
  effect on corrosion 162

PAH (polycyclic aromatic hydrocarbons) 5, 244
'Pepperpot' corrosion 70, 73
Permanganate
  as disinfectant 195

Index

pH
  influence of, on corrosion 158, 261
Phosphates
  to counter corrosion problems 164
Phosphoric acid
  to counter Pb corrosion 90
  fluxes 88
Pit depths
  in relation to time in Fe corrosion 236
Pitting, *see under* separate metals
Plastics fittings
  in stainless steel systems 110
Plastics pipes (*see also* Polymeric materials)
  in water supplies
    choice of material for 31
    corrosion of 32
    effect on water quality 40
    hydrolytic degradation of 32
    permeation of contaminants through 41
    use by the water industry 29
Plastic Pipes Club 140
Polyethylene sleeving
  use of 141
Polymeric materials
  absorption of chemicals by 274
  ageing of 276
  chlorinated PVC 29
  corrosion reactions with 274
  embrittlement of 274
  leaching of 10, 274
  oxidation of 274
  performance in water of
    polybutylene 29, 233
    polycarbonate 32
    polyester resins 33
    polyethylene 29, 37, 233
    polypropylene 233, 268
    PVC 40, 233
Polypropylene
  pipes, cracking of 268
  Polysaccharides
  occurence of, influence on corrosion 73, 225
Polyvinylidene difluoride 29

Positive lists (approved chemicals) 15
Potable water
  definition of 66
Pourbaix diagram 22
Pre-stressed concrete pipes 245
Prenormative research, needs for 13, 14, 16

Pseudomanas aeruginasa, see Copper and *Legionella*, and Microbially Induced Corrosion
PVCu
  introduction of and experience with 29, 245
  trials with 254

Quality (of water)
  Directorate General reference standard 249
  effects of, on corroison 115
  EC Directive 3
  public water supply
    UK 2
    USA 3
  WHO guideline 3

Red water 161, 262
Rehabilitation 252
Remedial work 252
Rulelearn
  automatic learning package 179

SEM (scanning electon microscopy)
  of corrosion films on copper 127, 225
  of corrosion products on steel 214
Scotland
  corrosion problems in water supplies in 59, 140, 224
Silicates to counter red water 164
Silt, *see* Deposits
Site
  anti-corrosion measures on 117
Solders
  lead containing, contamination from 90
  lead-free 66, 93

Solvents
  effects of, on plastics pipe materials 39
Spun iron
  use of 242
Stabilisers
  in plastics
    calcium as 44
    lead as 16, 41, 44
Stainless steels
  corrosion performance of 99
  experience in use of 101
  fittings 98
  highly alloyed 105
  hot water cylinders 102, 110
  hot water systems 102
  leaching from 106
  standards for tubing in 95
  sterilising agents, effect of 105
  surface finish 105
  tube manufacture 98
  uses of 94
Standards
  for water quality 2
Steel pipe
  corrosion of 207
Sulphate ion
  effect on copper 122, 128
Sulphate reducing bacteria, *see* MIC
Sweden
  study of corrosion of copper in 259

Trunk mains
  burst frequency of 247
Tuberculation
  in relation to pipe diameter 177

Water
  Byelaws Advisory Service 9
  industry liaison groups, list of 255
  quality
    effect of, on corrosion requirements
      in Finland 234
    in Sweden 259
    in UK 248
  Supply (water Quality)Regulations 1989 (*see also* European Regulations) 7, 79
Water Act (1945) 2
Water Act (1989) 3, 249
Welds
  corrosion, 210, 237
WRC Fittings and Materials Listing 98
World Health Authority
  Guidelines for Lead concentrations in water 92

X-ray
  diffraction analysis
    of copper corrosion products 128, 132, 139
  microprobe analysis
    of corrosion products on steel 214

Zinc (*see also* galvanised steel)
  content of, in water, values for 4, 6
  corrosion of 54